T0258781

THE POLITICS OF PROXIMITY

Increasingly, everyday living and practices depend on how mobility (and immobility) is articulated through the ever-present influence of a range of physical and virtual infrastructures. This book focuses in particular on the 'political' dimension of mobility and immobility, which plays a key role in establishing patterns of proximity in real and virtual co-presence. Proximity is seen as the result of choices, negotiations and practices carried out in different settings.

Drawing from different literature streams (Sociology, Organization Studies and Science and Technology Studies), this book analyses patterns of mobility in relation to new possibilities of organizing space, time, and proximity to others. Different phenomena – from memorial sites to migration, from urban mobility to mobile work – are analysed, illustrating different types of proximity through mobility and immobility. In doing so, this book offers a cross-cultural and innovative theoretical framing of issues linked to mobility, through the link with immobility and proximity.

The Politics of Proximity

Mobility and Immobility in Practice

Edited by

GIUSEPPINA PELLEGRINO

University of Calabria, Italy

LONDON AND NEW YORK

First published 2011 by Ashgate Publishing

Published 2016 by Routledge
2 Park Square, Milton Park, Abingdon, Oxon OX14 4RN
711 Third Avenue, New York, NY 10017, USA

Routledge is an imprint of the Taylor & Francis Group, an informa business

British Library Cataloguing in Publication Data
The politics of proximity : mobility and immobility in
   practice. -- (Transport and society)
   1. Spatial behavior. 2. Sociology, Urban. 3. Industrial
   location--Social aspects.
   I. Series II. Pellegrino, Giuseppina, 1974-
   304.2'3-dc22

Library of Congress Cataloging-in-Publication Data
Pellegrino, Giuseppina, 1974-
   The politics of proximity : mobility and immobility in practice / by Giuseppina Pellegrino.
      p. cm. -- (Transport and society)
   Includes index.
   ISBN 978-0-7546-7766-6 (hardback) -- ISBN 978-0-7546-9514-1 (ebook)
   1. Social mobility. 2. Transportation--Social aspects. 3. Space--Social aspects. 4. Time-
-Social aspects. I. Title.
   HT612.P45 2011
   305.5'13--dc22

                                                                                          2010045782

ISBN 9780754677666 (hbk)

Contents

List of Figures

List of Tables

Notes on Contributors

Carmelo Buscema is an Assistant Professor in Political Sociology and member of the Doctoral School 'A.G. Frank' in the University of Calabria, Italy. His most relevant interests of research concern cognitive capitalism, socio-political impact of ICTs, contemporary migrations, changing forms of citizenship and sovereignty. He carried out research and fieldwork activities in Spain, Mexico, United States, Ecuador and Italy. Amongst his most recent publications: 'La rappresentazione del mondo nella "sfera generale". Tra spettacolo, vita quotidiana e nuova testualità', in Jedlowski, P. and Affuso, O. (eds) *La sfera pubblica. Il concetto e i suoi luoghi.* Cosenza: Pellegrini Editore, 2010, 170–98; 'La sineddoche migratoria. Braccia, non-persone, uomini e modello win-win', in Buscema, C. et al. (eds) *Frontiere migratorie. Governance della mobilità e trasformazioni della cittadinanza.* Roma: Aracne, 2009, 15–53; *Tempi e spazi della rivolta. Epistemologia critica delle soggettività (migranti) e dell'antagonismo ai tempi della governance e della finanziarizzazione.* Roma: Aracne, 2009.

Matteo Colleoni is a Researcher in Urban Sociology and Assistant Professor in Urban Policies at the University of Milan Bicocca. He is a member of IATUR (*International Association Time Use Research*), Mo.Ve (*International, Non Governmental, Permanent, Observatory on Sustainable Mobility in Metropolitan Areas*) and *Inter-University Centre of Research on Urban Time.* His main research topics are: mobility and accessibility, use of time, urban change and methods of social research. He is author of several essays on these topics: *Tempi sociali. Teorie e strumenti di analisi.* Roma: Carocci, 2004; 'Accessibility and Social Equity. A Study in the Metropolitan area of Milan', in Mo.Ve – Final Technical Report, 2006, 37–56; *La ricerca sociale sulla mobilità urbana. Metodo e risultati di indagine.* Milano: Cortina, 2008; *Per-correre la vita. Autonomia e mobilità delle persone disabili.* Milano: Angeli, 2008 and *La città: bisogni, desideri, diritti. Dimensioni spazio-temporali dell'esclusione urbana.* Milano: Angeli, 2009.

Kjell Engelbrekt is an Associate Professor at the Department of Political Science, Stockholm University, and Senior Lecturer at the Swedish National Defense College. He works in the interdisciplinary field of international relations and presently focuses on the political sociology of globalization and the security/ nationalism nexus. A recurrent interest is the intellectual history of European social science. Engelbrekt's most recent journal articles are featured in *The European Legacy* (2009), the *European Law Journal* (2010) and the *Encyclopedia of Political Science* (forthcoming). In 2010 Routledge published *Russia and*

Europe: Building Bridges, Digging Trenches which he co-edited and co-authored with Bertil Nygren.

Eva Gerharz is a Senior Researcher at the Faculty of Sociology, Bielefeld University in Germany. Currently, she acts as Junior Professor for Internationalisation and Development at the Ruhr-University Bochum. She studied sociology, political science and law and received her doctoral degree from the University of Bielefeld in 2007. Her main research interests are development sociology, activism and identity politics, globalization, transnational and translocal dynamics in South Asia. She is currently completing *The Politics of Reconstruction and Development in Sri Lanka* and is co-editor of *The Making of World Society: Perspectives from Transnational Research* (2008, together with Remus Anghel Gabriel, Gilberto Rescher and Monika Salzbrunn).

Laura Gherardi is currently Researcher at the Faculty of Sociology of U.C. Milan. She obtained a joint PhD in Sociology from École des Hautes Études en Sciences Sociales (Paris) and U.C. (Milan) with a thesis, published in French – La Mobilité Ambiguë (Editions Universitaires Européennes 2010) – and in Italian – Mobilità Ambigua (Mondadori 2011) – on the link between mobility and power. She has been Visiting Fellow at Cities' programme (London School of Economics) and co-author, with Mauro Magatti, of *The City of Flows* (Mondadori 2010).

Paola Jirón is a Chilean academic from the Institute of Housing at the Faculty of Architecture and Urbanism at the University of Chile. She has carried out extensive research, teaching and consultancy work in the areas of housing, urban quality of life, and urban daily mobility practices. She is currently the Coordinator of the Master Programme on Residential Habitat at the University of Chile and Responsible Researcher for the FONDECYT funded research project on Urban Daily Mobility and Social Exclusion in Chile (www.santiagosemueve.com). She holds a Bachelors Degree from Concordia University, Canada, an MSc from University College London, UK and a PhD in Urban and Regional Planning at the London School of Economics and Political Science.

Maria Cristina Marchetti is a researcher in Political Sociology at the Faculty of Political Sciences of the University of Rome La Sapienza; in 1998 she was awarded a PhD in 'Sociology of culture and political processes'. Her research activity has mainly focused on the analysis of social and cultural change in a complex society (*Il consenso nelle società complesse*, 2001; *Manuale di comunicazione, sociologia e cultura della moda. Moda e Società*, 2004; *Tempo spazio e società. La ridefinizione dell'esperienza collettiva*, with Donatella Pacelli, 2007). In the few last years she has been interested in the European integration process, focusing on the intercultural communication and on the role of civil society in European decision making (*Il processo di integrazione europea. Comunicazione* *interculturale e ruolo dei media*, 2006; *Democrazia e partecipazione nell'Unione europea*, 2009).

Chaim Noy is a Senior Lecturer in Communication at the Sapir College, Israel. His main fields of interest include performance studies, language and semiotics, mobility, masculinity, tourism, and qualitative/experimental research methods. His recent books include *Narrative Community: Voices of Israeli Backpackers* (Wayne State University Press, 2006). His recent and forthcoming articles include: 'Inhabiting the Family-Car: Children-Passengers and Parents-Drivers on the School Run'. *SEMIOTICA* (in press). 'On Driving a Car and Being a Family: A Reflexive Ethnography', in Vannini, P. (ed.) *Material Culture and Technology in Everyday Life: Ethnographic Approaches*. New York: Peter Lang Publishing, 2009, 101–13.

Giuseppina Pellegrino is a Lecturer in Sociology of Culture and Communication at the Faculty of Political Sciences, University of Calabria (Italy). Her background is in Communication Studies (BA), Management and Innovation (MA), Sociology and Science and Technology Studies (PhD). Her current research projects focus on the interrelation between mobility and technological infrastructures, mobile work and design of advanced technological systems. She has been Visiting Fellow at Edinburgh University (RCSS 2002), Lancaster University (CeMoRe 2007) and IAS-STS Graz (2008). Amongst her latest publications (2009): 'The Electronic Body between Fragmentation and Concentration. Tracing Symptoms for an Ethics of Multiplicity', in Nyiri, K. (ed.) *Engagement and Exposure. Mobile Communication and the Ethics of Social Networking*. Vienna: Passagen Verlag, 169–77; 'Learning from Emotions towards ICTs: Boundary Crossing and Barriers in Technology Appropriation', in Vincent, J. and Fortunati, L. (eds) *Electronic Emotion*. Oxford: Peter Lang, 207–30.

Foreword

by John Urry

Department of Sociology, Lancaster University

For move you must! 'Tis now the rage,
The law and fashion of the Age.

(Coleridge, S.T. [1824] 1912. *The Poems of Samuel Taylor Coleridge*,
London: Oxford University Press)

Sometimes it seems as if all the world is on the move. Criss-crossing the globe are the routeways of countless social groups intermittently encountering one another at places of proximity. They search out in real and electronic databases the next coach, message, plane, back of lorry, text, bus, lift, ferry, train, car, web site, wifi hot spot and so on. The scale of this travelling is immense. In 1800 people in the US travelled 50 metres a day – they now travel 50 kilometres a day. Today world citizens move 23 billion kilometres; by 2050 it is predicted that that figure will have increased fourfold to 106 billion if resources are unlimited. And this travel is all about being proximate with other people, with notable events and with distinct places. This marvellous new book explores many different sites and dimensions of such multiple, contested and fought for and fought against proximities.

This pattern of mainly but not entirely voluntary travelling is the largest ever peaceful movement of people across borders. Until 2008 such movement has shown little sign of substantially abating in the longer term even with September 11, Bali, Madrid and London bombings, pandemics and other global catastrophes. Being physically mobile in order to be proximate has become for both rich and even for some poor a 'way of life'. The Internet has simultaneously grown incredibly rapidly, faster than any previous technology. Also there are worldwide 2–3 billion mobile phones and related mobile devices whose use occurs in part while on the move, while getting to and enabling new proximities as well as various reunions.

Partly in response to all this theorists as well as more empirical analysts are now mobilizing a 'mobility turn', a different way of thinking through the character of economic, social and political relationships in the contemporary world as the chapters here interestingly explore. Such a turn is spreading in and through the social sciences, mobilizing analyses that have been historically static, fixed and concerned with predominantly a-spatial 'social structures'. The mobility turn is post-disciplinary, concerned with multiple ways in which economic and social life is performed and organized through time and across various spaces of movement and proximity. Analyses of the complex ways that social relations are 'stretched' across the globe are generating theories, research findings and methods that 'mobilize' or assemble analyses of social ordering achieved in part on the move and contingently as processes of flow and contingent proximity.

Overall these mobilities have been a black box for the social sciences, generally regarded as a neutral set of processes permitting forms of economic, social and political life that are explicable by other more causally powerful processes. Holidaymaking, walking, car driving, phoning, flying and so on are mainly ignored by the social sciences although they are manifestly significant within people's everyday lives. Further there is a minimization of the significance of such movement for the nature of work relations, family life, leisure, politics and protest, all of which entail temporary proximity.

Each intersecting 'mobility' presupposes a 'system' (in fact many systems). These systems make possible movement: they provide 'spaces of anticipation' that the journey can be made, that the message will get through, that the parcel will arrive, that the family group can be proximate. Systems permit predictable and relatively risk-free repetition of the movement in question. These systems include ticketing, oil supply, addresses, safety, protocols, station interchanges, web sites, docks, money transfer, inclusive tours, luggage storage, air traffic control, barcodes, bridges, timetables, surveillance and so on. As daily and weekly time-space patterns in the richer parts of the world are desynchronized from historical communities and place, so these multiple systems provide the means by which work and social life can get scheduled and rescheduled. Organizing 'co-presence' with key others (workmates, family, significant others, friends) within each day, week, year and so on becomes more demanding with this loss of collective 'localized' coordination. The greater the personalization of networks, the more important are systems to enable the array of meetings and intermittent proximities that are necessary for family, friendship and professional life to be lived at a distance.

The 'social practices' of modern life thus have come to involve regular and predictable long distance movement of people (including commuters, holidaymakers, and making family and friendship visits) and objects (including water and food). This long distance mobile world is moreover deeply dependent upon, and embedded into, abundant cheap oil. Most industrial, agricultural, commercial, domestic, and mobility systems are built around the plentiful supply of 'black gold'. Oil powers virtually all movement of people, materials, foodstuffs, and manufactured goods around the world, providing 95% of transportation energy. It is remarkably versatile, convenient and during the twentieth century cheap. Oil became vital to virtually everything done and especially to everything that moves on the planet. This oil-based infrastructure was a twentieth-century phenomenon with the US as the disproportionately high energy producing and consuming society.

However, oil is not simply plentiful and growing in supply; three to four barrels of oil are consumed for every one discovered. Worldwide the largest oilfields were discovered over half a century ago, with the peak of oil discovery being in the mid-1960s. Over the longer term energy will become increasingly expensive and there will be frequent shortages because of the fall in per capita availability. Many assess that there is not enough oil to fuel such worldwide systems of global mobility needing, with 'business as usual', to double by 2050.

Thus twentieth-century capitalism generated the most striking of contradictions. Its pervasive, mobile and promiscuous commodification involved utterly unprecedented levels of energy production and consumption, a high carbon society whose legacy we are beginning to reap. This contradiction could result in a widespread reversal of many of the systems that constitute capitalism as it turns into its own gravedigger. A 'carbon shift' is inevitable. In the twenty-first century capitalism seems to be unable to control those powers that it called up by the spells set in motion during the unprecedented high carbon twentieth century which reached its peak of global wastefulness within the neo-liberal period from the 1980s onwards.

So the mobile twentieth century has left some constrained opportunities for mobility in this century. The future of 'mobility' is central to deciphering the future of life itself which could develop in one of a number of directions. What we should not assume is that the twentieth-century mobile world, especially building on innovations from around 1840, will continue to be the principle of organization into this century. Some now argue that climate change, the build-up of toxic chemicals in the environment and oil shortages will in the twenty-first century hugely constrain the possibilities of reengineering future mobilities and energy uses so as to avoid a 'societal collapse' through internal contradictions.

Mobile lives and contingent proximities for millions could only be a short moment in human history. For a century or so the rich world went mad before its contradictions so kicked in that twenty-first-century humans and machines will have to deal with a much slower legacy as societies go into reverse. Mobile lives and those temporary moments of proximity with people, places and events may be a short if remarkable interlude in the history of humans and their surprising mobile machines.

Preface

This book came out of a need and a curiosity: the need to make sense of communication at a distance – not by chance, I come from a Communication Studies background – and the fascination travelling, corporeal mobility and going around have had in my personal life and biography over the last decade.

I deeply believe that what we choose to look at is part of ourselves, in diverse modalities, both 'positive' and 'negative': we research on things which make us happy and passionate or upset and doubtful, sometimes sceptical. So it is the question of proximity and mobility for me, and the call for papers I launched in 2008 at the IIS World Congress of Sociology held in Budapest, responded to such needs and feelings. We always *feel* the world around *while moving on*, and *move on* because of *feeling* the world.

The Politics of Proximity is, in this respect, the result of choices and chances, emerging from contingencies stemming from a chain of events.

As usual, this edited book is the result of a patient and supportive process made possible by all of the contributors. Without their compliance and collaboration, this would not exist as such.

I would like to thank all of them, the Publisher and especially Valerie Rose and Jude Chillman for the patience and comprehension they always showed towards my requests and delays.

I owe many ideas of the introduction and inspiration for the research carried out over the last few years to John Urry, who welcomed me to CeMoRe as Visiting Fellow in Sociology in 2007.

A special thanks to my colleague and friend Ian Robinson, who made the non-native English of the authors smoother and easier for the (hopefully many) readers.

This book is dedicated to my family, and especially to my parents, for supporting me through continuous proximity and for bearing my compulsion to mobility and travelling around.

Giuseppina Pellegrino
Cosenza

No map there, nor guide,
no voice sounding, nor touch of human hand,
nor face with blooming flesh, nor lips, nor eyes, are in
that land.

(Darest Thou Now O Soul, Walt Whitman)

Stranger, if you passing meet me and desire to speak to
me, why should not I speak to you?
And why should you not speak to me?

(To You, Walt Whitman)

Introduction:
Studying (Im)mobility through a Politics of Proximity

Giuseppina Pellegrino

From John Urry's seminal book *Sociology beyond Societies. Mobilities for the 21st Century* (Urry 2000) onwards, this last decade has been marked by a novel interest towards mobility, mobilities, movement and motion as the hallmark of both social dynamics and their sociological (but not only sociological) understanding.

It is not by chance that the International Sociological Association (ISA) decided to entitle its 2010 World Congress 'Sociology on the Move'; neither by chance has the body of literature having mobility/mobilities as keywords and topic been increasing (Cresswell 2006, Sheller and Urry 2006a, Urry 2007, Canzler, Kaufman and Kesselring 2008, Dennis and Urry 2009, Verstraete 2009, Adey 2010, Cwerner, Kesserling and Urry 2009, Schönfelder and Axhausen 2010 amongst many others).

Is mobility a new phenomenon in history and society? Migrations, diasporas, urbanization processes show how big physical movements of people constituted turning points for creating new social, cultural, economic and political conditions. What changes, indeed, when looking at contemporary society by highlighting the continuous and interrelated movement of multiple elements (not only people, but also objects, information, representations, risks, cultures and communication) is the epistemological primacy so far attributed to sedentariness, physical proximity and stability (cf. Engelbrekt in this book).

The Simmelian matrix of such a perspective relies not only in the primacy of the sense of the eye for corporeal travel (cf. Urry 2000, 2007), but also in the relevance of relations and relationality, more than their provisional outcomes. In a certain sense, it could be said no sociology of change can exist, since sociology (as well as society) *is* change, movement and motion. Simmel also contributes greatly to an analysis of travel, body in motion, patterns of mobility in the city (Urry 2007).

Given this scenario where multiple forms of mobility are intertwined to the extent of producing complex patterns of real and virtual co-presence, what is the contribution that this book aims to provide in the context of the (over)crowded, ever changing landscape of mobility as academic discourse?

Indeed, mobility and related concepts (flow, flux, networks, -scapes, cf. Hannerz 1992, 2002, Castells 1996, Appadurai 1996) has constituted a consolidated discourse unifying some social sciences, from sociology to anthropology and cultural geography, over the last decade. In this sense, mobility is a powerful key-word used to assemble interpretive resources in order to understand the global changes occurring in contemporary society (cf. Hannam, Sheller and Urry 2006, Pellegrino 2007).

The title of the book summarizes its 'philosophy', based on a relational approach to the phenomena addressed, as multiple as the chapters and contributions herein collected: from theoretical analyses of space, mobility and proximity, within and beyond the so-called 'Mobility Turn' (Marchetti, Engelbrekt, Buscema); to issues of textual (im)mobilities linked to car transport system (Noy) and a return to physical proximity after segregation (Gerharz); from the city as landscape enabling and constraining mobility, immobility and inequalities (Colleoni, Jirón); to the global phenomenon of multinational travelling workforces (Gherardi).

What ties together so many different subjects, topics and focuses?

In the beginning, many of the chapters of this book came out of a call for papers for a session held at the IIS Congress of Sociology in Budapest (2008), starting from the evidence that intersections, overlaps and relations between globality and locality can be framed through the encompassing concept of mobility, which fosters both a powerful discourse in multiple settings and a renewed perspective in looking at socio-political transformations in the twenty-first century.

After that and through the complex process of peer reviewing and adaptation which allows for ideas to circulate in the form of what is known as an 'edited book', some key words and statements emerged as follows, constituting the (hidden) infrastructure of the book:

- the inescapably political character of proximity and, complementarily, mobility. 'Political' means here that more or less freely chosen (or constrained) elements are mobilized together when people, objects, information are on the move. In this respect, *politics* can be re-framed as the 'art', the power, and the possibility to set up strategies in order to enable, constrain or even enforce conditions of physical and virtual proximity between people, objects and information, or, in other terms, sociotechnical assemblies, networks of human and non human elements (Latour 1989);
- the need to put *proximity* at the centre of the stage, in order to understand how the feeling of being close to each other is accomplished through circulating objects, practices and discourses, ever more crucial in our social being and social life. Proximity itself is on the move and constitutive of mobility;
- the relational, ever changing, dialectic constitution of mobility (cf. Urry 2007, Adey 2006, 2009) which is addressed in this book through the conceptual couple *Mobility/Immobility*: neither a dualism nor an opposition, rather a relational *continuum*;

- the *sociotechnical constitution* of our everyday life as invisible, tacit *fil rouge* of all the contributions. None of them addresses explicitly the issue of technological innovation (Buscema being a partial exception); notwithstanding this apparent marginality of technology, it is in the backstage of many of the essays. This introduction will provide an overall view of the elements of the current technological scenario and its relevance for proximity, mobility and immobility;
- *practice* as the situated and material *locus* where proximity, mobility and immobility are put forward, challenged and realized, throughout myriad contexts, cases, situations and conditions where different assemblies of (im)mobility and types of proximity are practised and constructed.

The remainder of this introduction will provide a focused analysis of the categories listed above, in order to propose a general frame, through which to enter the book contents.

Politics

Politics is about choices and decisions, selection and norms. Politics is about rhetoric and persuasion: outcomes of decisions are not natural, but social; neither taken for granted, nor irreversible, as they are politically relevant and inspired by political moves and motives.

Since politics is about positioning oneself, situating selves and things 'somewhere' (Haraway 1991, Suchman 2002), as well as articulating the engineering of the heterogeneous which mingles together people, artefacts and words (Law 1997), it is involved in proximity and, consequently, in mobility.

In fact, '… Mobility provides a space for a politics and renders our ability to be *political* by shaping one's capacity to contest, deliberate and oppose' (Adey 2010: 84, original emphasis).

Furthermore, 'Mobilities are underscored by political decision making and ideological meanings that arrange mobility and the possibility of mobility – motility – in particular ways to relations of society and power' (Adey 2010: 131).

Giving proximity a political meaning aims to render the chosen, not natural, socially constructed and articulated character of the feelings of togetherness, vicinity and distance. The common association of mobility to freedom, equality and justice is the ideological veil which endows such a construction, the shortcomings of which can be more than severe, opening gates not just to a heaven of 'elected mobility', but also to the hell of injustice, inequalities and multiple divides (Adey 2010, Urry 2007). In this sense, proximity politics brings to the stage those issues linked with the forced relational closeness due to globalization processes, 'both "structurally" via the complex institutional interconnections of globalization, and "phenomenologically" via the sort of *experienced* proximity that is provided in time-space bridging technologies – particularly communications and media technologies' (Tomlison 2000: 403, original emphasis).

There are negotiations and conflicts, alliances and rivalries involved in this character of proximity. Proximity can be enforced or prohibited (cf. Buscema's and Gerharz's chapters in this book); it is based on communicational and metacommunicational skills, which can be acquired depending on social, cultural and mobility capital (cf. Part III of the book). Such a composite capital is the result of politics (and connected policies) which enable or constrain individuals and groups' ability to build up resources for accomplishing strategies of proximity and mobility towards each other.

Eventually, politics means selective decision making processes, and orientation towards future outcomes of present actions which are likely to happen (cf. Luhmann 1982). Such an orientation involves a proactive attitude, taking into consideration how the combination of multiple mobility systems generates potential for future mobilities, so shaping and changing patterns of (intermittent) proximity.

Proximity

The crucial character of proximity – and distance as its complementary dimension – is evident across the history of sociological thought and the media. The primacy of face-to-face, body-to-body relationships is maintained for the existence of both the individual and the community (cf. Gerharz in this book). Therefore, it can be said that:

> Social science presumes a 'metaphysics of presence', that it is the immediate presence with others that is the basis of social existence … And yet … all social life … presumes relationships of intermittent presence and modes of absence depending in part upon the multiple technologies of travel and communications that move objects, people, ideas, images across varying distances. (Urry 2007: 47)

Phenomenology approaches such a metaphysics and condenses it when stating that:

> The place which my body occupies within the world, my actual Here, is the starting point from which I take my bearing in space. It is, so to speak, the center 0 of my system of coordinates … And in a similar way my actual Now is the origin of all the time perspectives under which I organize the events within the world. (Schutz 1945: 545)

The immediacy and relevance of the 'Here and Now', or world of the 'actual reach' according to Schutz, presumes the possibility to derive from such a world all the other forms of communication and social interaction (cf. Berger and

Luckmann 1967). To some extent, the intervention of technical means aimed at extending spatial and temporal accessibility of symbolic contents, and producing what has been called 'mediated interaction' and 'mediated quasi-interaction' (Thompson 1995) represent surrogated forms of the original co-present, spatio-temporal simultaneity of face-to-face interaction. Yet, it is questionable if body-to-body communication must be considered as the only possible and univocal model for understanding mediated interaction (Fortunati 2005). Indeed, alternative models should consider that all types of interaction at a distance, intermittently and discontinuously performed, deserve the same attention and interest for social life as face-to-face communication based in physical co-presence (Urry 2002). At the same time, notwithstanding the issue of disruption of social ties linked to extensive networked communication being a prominent argument (cf. Adey 2010), compulsion to proximity in physical co-presence is not excluded, but rather enhanced, by imaginative, virtual and communicative travel of information (Urry 2002, 2007, Boden and Molotch 1994, Engelbrekt in this book).

Indeed, proximity, closeness and togetherness increasingly depend on how mobility is articulated through the ever-present influence of infrastructures: 'What constitutes social life is fundamentally heterogeneous and part of that heterogeneity … are various material objects … that directly or indirectly move or block the movement of objects, people and information' (Urry 2007: 50). This raises some important questions, variously addressed in the book's chapters:

- Is mobility a resource or a boundary (cf. Marchetti, Gherardi)?
- How is 'being on the move' accomplished (cf. Buscema, Noy)? How is the sense of time, space, global and local shaped through mobile practices (cf. Part I and II)?
- How is the sense of proximity constructed through multiple informational and communicational infrastructures (cf. Part II and III)?
- How are practices of mobility/immobility supported and fostered in the global arena (cf. Part III)?
- What is the relationship between proximity and contextualization (cf. Part I)?

(Im)mobility

Mobility cannot be conceived of without its opposite, that means immobility. Like Janus, the God of all beginnings, passages and movements, mobility has a double status and a relational constitution: '… mobilities are positioned in relation to something or somebody … mobility and immobility are understood as an effect or an outcome of a relation' (Adey 2010: 17–18). The category of difference becomes relevant here, with its political implications in terms of power and positioning into power relations (Butler 1990, Adey 2010; cf. Buscema and Part II of this book).

Difference makes it possible to overcome the idea of 'mobility' as single, rigid category, since it requires multiple moorings and relative immobilities for mobility to be maintained and performed over time and space in a fluid way (Urry 2003, 2007, Adey 2006, 2010).

The dialectic relationship between mobility and immobility, therefore, opens up their plurality and multiplicity: mobility is multiple, constituted by mobilities of various types (cf. Urry 2007), depending on the combination and action of different 'carriers' of movement (bodies, objects, goods, information, representations and so on). Furthermore, mobility and immobility rely on mediations and connections of such carriers in broader sets, or textures; what Urry, again, names as mobility-systems '... that distribute people, activities and objects in and through time-space and are key in the metabolic relationship of human societies with nature' (Urry 2007: 51).

Again, these systems have a political relevance and significance, by means of '... the effect of producing substantial inequalities between places and between people in terms of their location and access to these mobility-systems' (Urry 2007: 51). In this sense, '... unforced movement is power ... [It means] to be able to move (or to be able voluntarily to stay still) is for individuals and groups a major source of advantage' (Urry 2007: 51–2; cf. Gerharz and Gherardi in this book).

These systems are based on expert knowledge, therefore (im)mobility is deeply embedded and rooted in what I term as sociotechnical mediation, which accounts for the different and diverse meanings mobilities have in the twenty-first century society: '... while it is true that all societies have involved multiple mobilities ... the twenty-first century places *digitized* systems of mobility at its very core' (Urry 2007: 15, my emphasis). These systems and their qualification as 'digitized' constitute the novel, emergent and crucial character of contemporary mobilities. In the next section, a sociotechnical, non deterministic approach to qualities of these digitized systems will be proposed.

Sociotechnical mediation of (im)mobility

Far from being something external impacting unilaterally on our daily lives, technology is the result of conflict and negotiation among key social groups, which construct it (Bijker 1995). In particular, technology is deeply involved in the way people, objects and information are more and more 'on the move'. Mobility depends on sociotechnical processes which make artefacts increasingly convergent, multi-functional and pocketable. Technological mediation of mobility is both based on specific artefacts, e.g. mobile phones, laptops, PDAs; and embedded in complex infrastructures based on sociotechnical networks, e.g. electricity, the Internet, broadband networks, wireless networks. Such infrastructures are the invisible and embedded texture ('moorings', Urry 2003) which make it possible for people, objects and information to be mobile. They represent the pre-requisite of interconnections which allow communication while being on the move, as well as the portability and transferability of data and information across large networks.

Technologies of/for mobility can be situated at the crossroads of complementary phenomena which characterize contemporary sociotechnical mediation, namely convergence, or the trend towards uniformity of technological platforms and systems; saturation; hybridity; ubiquity: saturation as the web of interoperability on which infrastructures are built up and linked to each other; hybridity as the constant interlinkage of human and non human components; ubiquity as aspiration towards omnipresence through simultaneity and instantaneity (cf. Pellegrino 2010a).

Convergence in technologies for mobility

Different types of convergence can be identified in technologies for mobility. Generally speaking, technologies tend to converge at the level of markets, functions and infrastructural architectures. Furthermore, there is a material profile of this convergence, which has to do with miniaturization and portability of multiple, multifunctional mobile technological artefacts. They concentrate in themselves a high diversity of tasks, functions and channels of communication. Multimedia is part of the emergence of convergence as a long-lasting trend in media and information history. Such a convergence is particularly linked with the body and redefines materiality and visibility of technology. The increasing convergence of functions and infrastructures for corporeal and communication travel (e.g. the smart car and the development of the mobile Internet) is also part of the sociotechnical framework of convergence.

Saturation of mediated environments

The texture of saturation is, like that of infrastructures, integrated and based on the concept of interoperability (Bowker and Star 2000). As such, it is invisible, transparent, therefore difficult to grasp in its patterns. It happens with all complex infrastructures that we become aware of their existence when they stop working, when any kind of breakdown, interruption and misuse occurs. This phenomenon can be referred to the mobile phone as ubiquitous technology accessible everywhere/everytime, the saturation of which increases expectations of continuous availability of participants to the communication process (Katz and Aakhus 2002). More generally, the concept of saturation describes well the way our bodies and environments are intertwined into inextricable chains of socio-technical relationships, like in the 'everyware' texture of ubiquitous computing, imagined as a technology able to colonize surfaces and settings of everyday life (Greenfield 2006).

Hybrid bodies and artificial natures

Objects ready to hand which provide affordances to mobile practices (Urry 2007) are differently integrated, mingled and portable within the body and the surrounding environment. Interoperability and saturation make technological devices, networks and media closer to each other. In this respect, as they become closer, differences and boundaries between them and between technology and ourselves go in the background. Indeed, proliferation of hybrids and the erection of the boundary between nature and culture are part of the modernization process (Latour 1993). Everyday, we delegate our actions and perform activities through some sociotechnical device, so that it becomes an integral part of our sociality and inextricably assembled with our agency. What appears to be 'natural' is highly artificial and artificially naturalized through rhetorical strategies. All of us are hybrid and hybridized: the body is increasingly empowered through technology; communication cannot be conceived of without mixtures of media and assemblies of sociotechnical devices.

Ubiquity and extension of co-presence

Ubiquity can be defined as the tension towards 'being anywhere anytime' as opposed to the *hic et nunc* constraints of face-to-face interaction. The mobile phone is *par excellence* an example of such a ubiquity because of the constant availability it makes possible. The tension towards reaching a virtual, potential omnipresence is supported by convergent artefacts, which make ubiquity more at hand than before. Mobility itself is a ubiquitous phenomenon, and globalization can be interpreted as a process of extensive mobilities (Adey 2010).

Ubiquity as aspiration to omnipresence is embedded into discourses, information and artefacts supposed to be accessible anywhere anytime (at least in principle). The myth of ubiquitous computing as invisible, unobtrusive infrastructure embedded into material surfaces is exemplary of a trend to imagine and design contexts of interaction, both public and private, where materiality of technology and transparency of infrastructure are redefined, so mobilizing resources to build up future social settings (cf. Pellegrino 2010b).

Practice

Practice is usually contrasted with theory, since it privileges the situated *locus* of action as the focus of interest in understanding social life and relations. Drawing on phenomenological accounts (Merleau-Ponty 1962, Schutz 1945, Berger and Luckmann 1967), especially Organization Studies (Gherardi 2001, 2006, Orlikowski 2000) as well as Science and Technology Studies (Suchman 1987, Law 1997, Suchman et al. 1999) have pointed out how practice is the foundation for an alternative understanding of organizations and technologies, and in general for human cognition and activity (cf. Engeström and Middleton 1996). Such an alternative understanding, or the 'added value' of a practice perspective, relies on the fact that:

> it enables analysis of the social connections among individuals, collectives, organizations, institutions, the situated contexts in which these connections take specific form, and all the intermediaries utilized by them (…) dynamically as the constant becoming of a form which self-reproduces but is never identical with itself in that practices are incomplete and indeterminate until they are situatedly performed. (Gherardi 2006: XVIII)

Such a definition of practice as 'connection in action' or 'texture of practice' elaborated by Silvia Gherardi (Gherardi 2006) seems particularly adequate to study (im)mobility. Not least because it contains a Simmelian echo of the struggle between form and life, object and process, subjective and objective spirit. On other hand, it is coherent with the attention the 'Mobilities Paradigm' has devoted to complexity theory (Urry 2003, 2007), looking at patterns of self production and maintenance of mobility systems and their fluidity. Therefore, it can be said that practice shapes (im)mobility as well as proximity through multiple, situated *loci* of temporary stabilization and fluidity, where a tentative trajectory for assemblages of intermediaries is established and accomplished.

Practice is one of the key-words of this book, not only in the senses listed above, but also because of the attention reserved for seeing mobility and moving empirically (six of the eight chapters are based on empirical research). Doing mobility and enacting immobility, in other words looking at the situated contexts where mobility, immobility and proximity are displayed, accomplished and realized, is one of the objectives of this book. Unfolding representations and discourses of (mediated) (im)mobility is possible through a practice-based approach, which involves the awareness that 'humans are sensuous, corporeal, technologically extended and *mobile* beings' (Urry 2007: 51, original emphasis) as well as the necessity to look at the body (and not only it) *in action*, as indivisible from the mind, and embedded in the surrounding environment (cf. Merleau-Ponty 1962, Adey 2010). Following the actors, or better the hybrid, heterogeneous actants in their networks of alliances, enrolment and fabrication (Latour 1989, 1996) as *in fieri* processes, means to avoid the risk of reifying the object. This is even more important when dealing with mobility, which is by definition a changing category.

Last but not least, a practice based perspective recognizes the impossibility of telling and saying everything, the necessity of 'doing' the experience of moving on, and the incommensurability between formal and informal settings, explicit and tacit knowledge (Brown and Duguid 1998, Orr 1996, Polanyi 1966).

Practice goes back to situating things, people, connections in action: it complements the deep political meaning of proximity and (im)mobility, as well as the attempt to understand them while making their constitutive, complex multiplicity as accountable as possible.

The book's contribution – an outline

Given the scenario depicted above, this book aims to focus on specific dimensions of the 'multiple mobilities paradigm' (Sheller and Urry 2006b, Urry 2007), putting forward the following hypotheses to understand contemporary mobility:

- the 'political' dimension as constitutive in establishing patterns of proximity in real and virtual co-presence (cf. Urry 2000). As a consequence, (the sense of) proximity is seen as the result of choices, negotiations and practices carried out in different settings, specified and presented along the various chapters;
- the irreducibly relational character of mobility (Adey 2006) which can be understood only and always in function of its oppositional constituency (that means, immobility). For objects, people and representations being mobile, others must be immobile, anchored, embedded. Such a relational quality is particularly important in shaping mobility and the sense of proximity itself;
- the central concept of 'practice' in shaping and situating relationships between proximity, mobility, and immobility.

The main scope of the book is, therefore, to analyse patterns of mobility in relation with new possibilities to organize space, time and proximity to others.

The main features of the book are the following:

- wide range of methodologies, approaches and case studies through eight chapters;
- equal attention to the theoretical, empirical and methodological dimension of mobility/proximity;
- different phenomena analysed as creating different types of proximity through mobility and immobility (from automobility to diaspora, from urban mobility to mobile work).

The objective is to offer a cross culture (case studies cover a wide range of countries) framing of issues linked to mobility, through the link with immobility and proximity.

This introductory chapter has focused on the points above to specify the contribution the book aims to give to the current debate in the field of the so-called 'Mobility Turn' in sociology and to the vast sociological debate about modes of co-presence and their technological mediations.

Consistent with this analysis, the book is structured according to three conceptual axes, each constituting a different part of the whole work.

Part I (*Categories of Proximity/Mobility*) is conceived of as the most theoretically grounded part, aimed at framing:

- the categories of space, time and place in the vast sociological debate (Marchetti's contribution);
- a critical phenomenological perspective as an attempt to enrich Urry's 'Mobilities Paradigm' (Engelbrekt's chapter);
- the notion of proximity/mobility in the framework of a critique to capitalist society, and its impact on the organization of work (Buscema's work).

In the second part of the book (*Discourse/Identity in Proximity and Mobility*), two notions are central: 'discourse', as in the semiotical, discursive analysis of two cases in the automobility system (Noy's chapter on discursive automobilities in Israel); and 'identity' as emerging from enforced proximity or isolation (Gerharz's analysis of Jaffna re-migration patterns).

Part III (*Global Firms/Urban Landscapes as Scenery for Proximity and Mobility*) proposes two different contexts to analyse current scenarios of proximity and mobility, emphasizing the danger of inequalities, asymmetries and human costs in accessing mobility as a resource. The multinational company is the setting of Gherardi's study of a peculiar mobile group (international managers). Urban landscapes (four European cities, in Colleoni's chapter; Santiago de Chile in Jirón's qualitative study) constitute another relevant arena where proximity and mobility are performed, resulting in different outcomes in terms of inequalities, borders and accessibility to (im)mobility resources.

Concluding remarks

The introductory chapter of this book argues that proximity, and social relationships stemming from it, embody political meanings, which are performed through extensive sociotechnical systems of mobility and immobility. These systems can both enable and constrain the movement of people, objects, information and representations across distance, at the same time allowing for the understanding the specificity and peculiarity of their trajectories of motion and resilience in situated contexts.

The eight chapters of the book, divided into three parts, represent an articulation of this general thesis, focusing on theoretical aspects of the relationship between mobility, immobility and proximity (Part I); on discursive practices and identity shortcomings of proximity through forced mobility (Part II); on the local-global nexus of mobility and immobility (Part III).

(Im)mobility is, first and foremost, a relation to the world which shapes our sense of closeness and distance to people, objects, ideas and information, allowing us to feel and perceive ourselves through both permanence and instability in space, time and society. The peculiar forms this relation can take place in across cultures, contexts, and practices, are the voices which compose this book. The list is, as usual, very tentative, as things (and beings) always *move on*.

References

Adey, P. 2006. If mobility is everything then it is nothing: towards a relational politics of (im)mobilities. *Mobilities*, 1(1), 75–94.

Adey, P. 2010. *Mobility*. London: Routledge.

Appadurai, A. 1996. *Modernity at Large. Cultural Dimensions of Globalization*. Minneapolis, London: University of Minnesota Press.

Berger, P.L. and Luckmann, T. 1967. *The Social Construction of Reality*. New York: Doubleday.

Bijker, W.E. 1995. *Of Bicycles, Bakelites and Bulbs*. Cambridge, MA: The MIT Press.

Boden, D. and Molotch, H. 1994. The compulsion to proximity, in *Nowhere. Space, Time and Modernity*, edited by R. Friedland and D. Boden. Berkeley: University of California Press.

Bowker, G. and Star, S.L. 2000. *Sorting Things Out. Classification and Its Consequences*. Cambridge, MA: The MIT Press.

Brown, J. and Duguid, P. 1998. Organizing knowledge. *California Management Review*, 40(3), 90–111.

Butler, J. 1990. *Gender Trouble: Feminism and the Subversion of Identity*. London: Routledge.

Canzler, W., Kaufmann, V. and Kesselring, S. (eds) 2008. *Tracing Mobilities. Towards a Cosmopolitan Perspective*. Aldershot: Ashgate.

Castells, M. 1996. *The Rise of the Network Society*. Oxford, New York: Blackwell.

Cresswell, T. 2006. *On the Move: The Politics of Mobility in the Modern West*. London: Routledge.

Cwerner, S., Kesselring, S. and Urry, J. (eds) 2009. *Aeromobilities*. London: Routledge.

Dennis, K. and Urry, J. 2009. *After the Car*. Cambridge: Polity Press.

Engeström, Y. and Middleton, D. (eds) 1996. *Cognition and Communication at Work*. Cambridge: Cambridge University Press.

Fortunati, L. 2005. Is body-to-body communication still the prototype? *The Information Society*, 21(1), 1–9.

Gherardi, S. 2001. From organizational learning to practice-based knowing. *Human Relations*, 54(1), 131–9.

Gherardi, S. 2006. *Organizational Knowledge: The Texture of Workplace Learning*. Oxford, Malden, MA, Carlton: Blackwell Publishing.

Greenfield, A. 2006. *Everyware. The Dawning Age of Ubiquitous Computing.* Berkeley, CA: New Riders.

Hannam, K., Sheller, M. and Urry, J. 2006. Editorial: mobilities, immobilities and moorings. *Mobilities*, 1(1), 1–22.

Hannerz, U. 1992. *Cultural Complexity. Studies in the Social Organization of Meaning.* New York: Columbia University Press.

Hannerz, U. 2002. Flows, boundaries and hybrids: keywords in transnational anthropology. [online] Available at: www.transcomm.ox.ac.uk/working%20papers/hannerz.pdf [accessed: 30 June 2006].

Haraway, D. 1991. Situated knowledges: the science question in feminism and the privilege of partial perspective, in *Simians, Cyborgs, and Women*, edited by D. Haraway. New York: Routledge, 183–201.

Katz, J.E. and Aakhus, M. (eds) 2002. *Perpetual Contact. Mobile Communication, Private Talk, Public Performance.* Cambridge: Cambridge University Press.

Latour, B. 1989. *Science in Action.* Oxford: Oxford University Press.

Latour, B. 1993. *We Have Never Been Modern* (transl. by Catherine Porter). Cambridge: Harvard University Press.

Latour, B. 1996. *Aramis or the Love of Technology*, Cambridge MA: Harvard University Press.

Law, J. 1997. Heterogeneities. [online] Available at: http://www.lancs.ac.uk/fass/sociology/papers/law-heterogeneities.pdf [accessed: 20 May 2006].

Luhmann, N. 1982. *The Differentiation of Society.* New York: Columbia University Press.

Merleau-Ponty, M. 1962. *Phenomenology of Perception.* London: Routledge and Kegan Paul.

Orlikowski, W.J. 2000. Using technology and constituting structures: a practice lens for studying technology in organizations. *Organization Science*, 11(4), 404–28.

Orr, J.E. 1996. *Talking About Machines: An Ethnography of a Modern Job.* Ithaca, NY: ILR Press.

Pellegrino, G. 2007. Discourses on mobility and technological mediation: the texture of ubiquitous interaction. *PsychNology Journal*, 5(1), 59–81.

Pellegrino, G. 2010a. Mediated bodies in saturated environments. Participation as co-construction, in *Interacting with Broadband Society*, edited by L. Fortunati et al. Berlin: Peter Lang, 93–105.

Pellegrino, G. 2010b. *Future's on the Move. Looking Forward to Ubiquitous Communication?* Paper to XVII World Congress of Sociology: Sociology on the Move, Gothenburg, 11–17 July.

Polanyi, M. 1966. *The Tacit Dimension.* New York: Doubleday.

Schutz, A. 1945. On multiple realities. *Phenomenology and Social Research*, 5(4), 533–76.

Schönfelder, S. and Axhausen, K.W. 2010. *Urban Rhythms and Travel Behaviour: Spatial and Temporal Phenomena of Daily Travel.* Aldershot: Ashgate.

Sheller, M. and Urry, J. 2006a. *Mobile Technologies of the City*. London, New York: Routledge.

Sheller, M. and Urry, J. 2006b. The new mobilities paradigm. *Environment and Planning A*, 38(2), 207–26.

Suchman, L.A. 1987. *Plans and Situated Actions*. Cambridge: Cambridge University Press.

Suchman, L.A. 2002. Located accountabilities in technology production. *Scandinavian Journal of Information Systems*, 14(2), 91–105.

Suchman, L.A., Blomberg, J., Orr, J. and Trigg, R. 1999. Reconstructing technologies as social practice. *American Behavioral Scientist*, 43(3), 392–408.

Thompson, J.B. (1995). *The Media and Modernity*: *A Social Theory of the Media*. Cambridge: Polity Press.

Tomlinson, J. 2000. Proximity politics. *Information, Communication & Society*, 3(3), 402–14.

Urry, J. 2000. *Sociology beyond Society: Mobilities for the 21st Century*. London: Routledge.

Urry, J. 2002. Mobility and proximity. *Sociology*, 36(2), 255–74.

Urry, J. 2003. *Global Complexity*. Cambridge: Polity Press.

Urry, J. 2007. *Mobilities*. Cambridge: Polity Press.

Verstraete, G. 2009. *Tracking Europe: Mobility, Diaspora, and the Politics of Location*. Durham: Duke University Press.

PART I
Categories of Proximity/Mobility

Chapter 1

Space, Mobility and New Boundaries: The Redefinition of Social Action

Maria Cristina Marchetti

Introduction

The increasing levels of mobility, which encompass globalization, migrations, tourism, media and technologies have implied a redefinition of space as a frame for social action.

In the sociological debate there are two prevailing notions of space: on the one hand there is the *physical space*, geometric, material, which equates the *space* with the *place*; on the other, the *social space*, abstract, immaterial, which is strictly related to the *space of interaction*. Classical sociology has unified both the notions, focusing on the analysis of social interactions within a specific place (*face-to-face* model).

However, the identification of *social* and *physical space* is now questioned from different points of view now overcome on different issues:

1. a shared, physical space does not entail a social space (Augé's 'non-places');
2. on the contrary, a not shared physical space, does not necessarily entail the lack of a social space (Meyrowitz, 'no sense of place').

These opposite processes imply the 'dematerialization' of space; this comes as a result of new ideas about mobility (Urry's 'new mobilities paradigm'; Castells's 'space of flows') and, as a consequence, the redefinition of social action: should people live without boundaries? Are there new boundaries emerging (the space of local; the space of culture)?

The notion of space in the sociological debate

Physical space vs. social space

The notion of space, together with time and culture, is currently one of the main fields to verify a long-term process leading to a redefinition of social action (Pacelli and Marchetti 2007). Migratory processes, globalization and mass media have changed how space is conceived in social sciences. In the sociological debate, there are two prevailing notions of space: on the one hand we have the *physical space*, geometric, material, which equates the *space* with the *place*; on the other we have the *social space*, abstract, immaterial, which is strictly related to the *space of interaction*. It is important to point out the distinction between *place* and *space* because they are often used as synonymous:

> An essential preliminary here is the analysis of the notion of place and space suggested by Michel de Certeau. He himself does not oppose 'place' and 'space' in the way that 'place' is opposite to 'no place'. Space, for him, is a 'frequented place', 'an intersection of moving bodies': it is the pedestrian who transforms a street (geometrically defined as a place by town planners) into a space. This parallel between the place as an assembly of elements coexisting in a certain order and the space as animation of these places by the motion of a moving body is backed by several references. The first of these references is to Merleau Ponty who in his *Phenomenologie de la perception* draws a distinction between 'geometric space' and 'anthropological space' in the sense of 'existential' space, the scene of an experience of relations with the world in the part of a being essentially situated 'in relation to a milieu'. (Augé 1995: 80)

Similarly, in his essay on the *Consequences of Modernity*, Giddens says that:

> 'Place' is best conceptualised by means of the idea of locale, which refers to the physical settings of social activity as situated geographically. In pre-modern societies, space and place largely coincide, since the spatial dimensions of social life are, for most of the population, and in most respects, dominated by 'presence' – by localised activities. The advent of modernity increasingly tears space away from place by fostering relations between 'absent' others, locationally distant from any given situation of face-to-face interaction. In conditions of modernity, place becomes increasingly phantasmagoric: that is to say, locales are thoroughly penetrated by and shaped in terms of social influences quite distant from them. (Giddens 1990: 18–19)

Classical sociology has unified both the notions, focusing on the analysis of social interactions within a specific place. Consequently, social space and physical space coincide, as simplified in the *face-to-face* model. As Meyrowitz assesses,

> [t]he relationship between physical place and social situation still seems so natural that we continue to confuse physical places with the behaviours that go on in them. The words 'school' and 'home', for example, are used to refer both to physical buildings and to certain types of social interaction and behaviour. (Meyrowitz 1985: 116)

Since the end of the nineteenth century, classical sociology has focused on the relationship between physical and social space. At that time several authors realized that the changes which occurred in the organization of spatial forms in the contemporary societies implied changes in the way people constitute social relationships. The increasing *density* within the metropolis space, as Durkheim stressed, produced an acceleration of social interaction, from which a new form of solidarity was born (Durkheim 1984). Similarly, the metropolis became the symbol of the progressive objectivation and abstraction of social relationships (Simmel 1950).

The sociological debate on the notion of space can be summed up in two prevailing interpretative orientations.

On the one hand determinism, which characterized a conspicuous part of classical sociology, assessed the supremacy of physical over social space; according to this perspective, the way physical spaces are conceived influences the interaction processes. On the other, the relational approach reversed the link between physical and social space: the way individuals interact reshapes the physical space they live in. Goffman can be considered the leading figure of the former approach and Simmel of the latter.

Goffman analysed the relationship between physical and social space, also offering some new perspectives on this topic. The language that Goffman uses is taken from the theatre and has strong spatial implications: the front stage and the backstage define a framework for social action.

The limit of Goffman's analysis relies on the fact that he does not provide a comprehensive description of the relationship between different representations. On the contrary, the social actors can play their role on different stages, using metalanguages, which refer not only to the physical context, but also to past experiences, and images processed through past interactions, too.

As regards the relational approach, maybe Simmel was the classical author who could be said to have enlarged and to some extent overcome the traditional approach on space. As Frisby stresses 'space is dealt with by Simmel in a complex manner involving both boundaries, distance and the removal of boundaries (the money nexus overcoming spatial boundaries)' (Frisby 1992: 76).

According to Simmel, space is a soul activity that the individual uses to connect disjointed experiences. Space is the '*in-between*' which is filled up with a reciprocal action (relationship). He particularly focuses on five properties of space: 1) the exclusivity or uniqueness; 2) the boundary; 3) the localizing or fixing of social interaction in space; 4) the proximity/distance in space; 5) the mobility (the changing of location). For the specific purposes of this chapter we will focus on the notion of *boundary* and on the significance of *proximity/distance* in space. These concepts are relevant in every sociology of space.

The former implies that space can be framed in by boundaries. As Frisby says, Simmel 'indicates that a society, and forms of sociation, possess a sharply demarcated existential space in which the extensiveness of space coincides with the intensity of social relationship' (Frisby 1992: 105). The latter maybe is the main contribution of Simmel's analysis of space'. As Levine has stressed:

> nearly all of the social processes and social types treated by Simmel may be readily understood in terms of social distance. Domination and subordination, the aristocrat and the bourgeois, have to do with relations defined in terms of 'above' and 'below'. Secrecy, arbitration, the poor person, and the stranger are some of the topics related to the inside-outside dimension. (Levine in Frisby 1992: 107)

The distance can be considered as immaterial and as a protection from invasions coming from the outside, which occurs in the metropolis. In a more complex manner, we can seemingly simultaneously participate and distance ourselves, as the figure of the flâneur does.

Particular attention is due to the fifth property of space – mobility – the only one that implies the shift from a condition of stasis to a condition of movement. Mobility presumes that individuals move from one place to another and as a result they have to cope with the change of the spatial conditioning on their lives. Nomadism and migration are the main examples of a mobile notion of space.

In this context the 'stranger' becomes the symbol of a condition that includes either stasis or movement. 'If wandering is the liberation from every given point in space, and thus the conceptional opposite to fixation at such a point, the sociological form of the 'stranger' presents the unity, as it were, of these two characteristics' (Simmel 1950: 402).

The stranger lives in a 'no-man's land' divided between nearness and remoteness. 'The unity of nearness and remoteness involved in every human relation is organised, in the phenomenon of the stranger, in a way which may be most briefly formulated by saying that in the relationship to him, distance means that he, who is close by, is far, and strangeness means that he, who also is far, is actually near' (Simmel 1950: 402).

He lives here and elsewhere and he experiences an interior nomadism, close to postmodern condition. 'The stranger is by nature non "owner of soil" – soil not only in the physical, but also in the figurative sense of a life-substance which is fixed, if not in a point in space, at least in an ideal point of the social environment' (Simmel 1950: 403).

From 'no sense of place' to 'non-places'

Nowadays, the identification of *social* and *physical space* is overcome on different issues:

1. a shared, physical space does not entail a social space (Augé's 'non-places');

2. on the contrary, a not shared physical space, does not necessarily entail the lack of a social space (Meyrowitz, 'no sense of place').

The first issue is related to the concept of 'no-place' proposed by Marc Augé. As Augé suggests, 'If a place can be defined as relational, historical and concerned with identity, then a space which cannot be defined as relational, or historical, or concerned with identity will be a no-place' (Augé 1995: 77–8).

A 'no-place' is a sort of 'free zone' where people meet each other without interacting: stations, airports, commercial areas, refugee camps, amusement parks, post offices and banks are conceived through this characteristic. They are *physical* spaces, in which people walk or stay for a short period but do not interact. It is difficult to say if absolute 'non-places' could really exist or a tendency to personalization can emerge even within them. The individuals need to reappropriate space either through social interactions or personal adaptation (Rossi 2006).

The second change which has occurred in the last few decades is due to the growing importance of mass media. This influence is quite opposite the notion of 'non-places'.

The thesis is that the perfect correspondence between *physical* and *social* space, as described in the *face-to-face model*, belongs to the pre-modern societies, in which all the experiences were firsthand ones.

On the contrary, mass media determine a *sui generis* interaction space, which integrates different communication models, refusing any *sense of place*.

In this context, in his work *No Sense of Place*, Meyrowitz tries to mediate Goffman's *face-to-face* model and McLuhan's media communication model:

> Goffman offers one factor that models behaviour: the 'definition of the situation' as it is shaped by particular interactional settings and audiences. Yet Goffman explicitly ignores changes in roles and the social order. McLuhan, on the other hand, points to widescale change in social roles resulting from the use of electronic media, but he provides no clear explanation of how and why electronic media may bring about such change. (Meyrowitz 1985: 4)

The lowest common denominator between Goffman and McLuhan is the structure of 'social situations'.

> I suggest that the mechanism through which electronic media affect social behaviour is not a mystical sensory balance, but a very discernible rearrangement of the social stages on which we play our roles and a resulting change in our sense of 'appropriate behaviour'. For when audiences change, so do the social performances. (Meyrowitz 1985: 4)

Meyrowitz says that 'situations' are defined in terms of behaviours in physical places. Physical and social space are so rejoined; for example, when we say that a behaviour does not come up to somebody's expectations, we say that it is 'out of place'.

Similarly he says that:

> although there are many logical reasons for the traditional focus on the place-bound situations, a question that arises is whether behavioural settings must be places. That is, is it actually place that is a large determinant of behaviour, or is it something else that has traditionally been tied to, and therefore confused with, place? There is another key factor besides place mentioned in Goffman's definition of regions that tends to get lost in most of his and other situationists' discussions of behavioural settings: 'barriers to perceptions'. Indeed, a close examination of the dynamics of situations and behaviour suggests that place itself is actually a sub-category of this more inclusive notion of a perceptual field. For while situations are usually defined in terms of who is in what location, the implicit issue is actually the types of behaviour that are available for other people's scrutiny. (Meyrowitz 1985: 36)

The interaction context broadens and includes information that does not derive from the physical context. 'It is not the physical setting itself that determines the nature of interaction, but the patterns of information flow. Indeed, the discussion of the definition of the situation can be entirely removed from the issue of direct physical presence by focusing only on information access' (Meyrowitz 1985: 36). Meyrowitz stresses that '"Information" is used here in a special sense to mean *social* information: all that people are capable of knowing about the behaviour and actions of themselves and others. The term refers to that nebulous "stuff" we learn about each other in acts of communication' (Meyrowitz 1985: 37).

Mediated and *un*-mediated interactions are strictly related and sometimes overlapping. Thus Meyrowitz says that the meaning of space in communication processes must be reconsidered. 'As a result of electronically mediated interactions, the definition of situations and of behaviours is no longer determined by physical location. To be physically alone with someone is no longer necessarily to be socially alone with them' (Meyrowitz 1985: 117).

In everyday life, *face-to-face* interactions and mediated interactions are mixed together. This redefines the notion of 'space of interaction': it is free from any references to a specific context (place) and includes different ways of communication.

A not shared physical space, does not necessarily entail the lack of a social space, but simply the redefinition of social space, regardless of *face-to-face* interactions, or furthermore, including them in a broader set of mediated, un-mediated, and even imagined experiences.

Meyrowitz shows how dualism, which opposes different forms of interaction, is replaced by their overlap in everyday life. As he stresses, 'it is very difficult to respond to one situation as if it were another. Parents' exhortations that a child must finish all the food in a plate "because children are starving in Africa" are usually ineffective because so rarely are there any starving African children peering through the window' (Meyrowitz 1985: 40–41). Many events could influence our behaviour even if we have never been present at them.

At the present time, individuals act 'as if' they had all the information about the others, looking at past experiences, images, second-hand information, all elements that are able to influence the final result of such an interaction.

The 'dematerialization' of space

The 'space of flows'

The separation of physical and social space succeeded in producing a 'dematerialization' of space. A new form of space was born, defined by Castells as 'space of flows', which characterized the network society. 'The space of flows is the material organization of time-sharing social practices that work through flows' (Castells 2000: 442).

Contemporary societies are being constructed on flows of capital, flows of information, flows of technology, flows of organizational interaction, flows of images, sounds, and symbols. 'By flows I understand purposeful, repetitive, programmable sequences of exchange and interaction between physically disjointed positions held by social actors in the economic, political, and symbolic structures of society' (Castells 2000: 442).

The space of flows, as the material form of support of dominant processes and functions in the network society, can be described by the combination of different layers of material supports:

> *The first layer, the first material support of the space of flows, is actually constituted by a circuit of electronic exchanges* (...) that, together, form the material basis for the processes we have observed as being strategically crucial in the network society. This is indeed a material support of simultaneous practices. Thus, it is a spatial form, just as it could be 'the city' or 'the region' in the organization of merchant society or of the industrial society. (Castells 2000: 442, my emphasis)

> *The second layer of the space of flows is constituted by its nodes and hubs.* The space of flows is not placeless, although its structural logic is. It is based on an electronic network, but this network links up specific places, with well-defined social, cultural, physical, and functional characteristics. (Castells 2000: 443, my emphasis)

Location in the node links the locality with the whole network. Both nodes and hubs are hierarchically organized according to their relative weight in the network. Nevertheless, this hierarchy may change, as it depends upon the evolution of activities processed through the network. Some places may indeed be switched off the network, and their disconnection implies an instant decline, together with economic, social and physical deteriorations.

The functions to be fulfilled by each network characterize the places that become their privileged nodes. In some cases, the most unlikely places become central nodes because of their historical specificity. None of these localities can exist by itself in such a network. In the functional dimension of the network society Castells proposes the case of the city of Rochester (Minnesota) and the Parisian suburb of Villejuif that became central nodes of a world network of advanced health research.

A similar approach has driven the activity of the Guggenheim Foundation in the promotion of contemporary art. What is the link between New York, Venice, Bilbao, Berlin and Dubai? Regardless of their location, they are central nodes of a global network of culture promoted by the Foundation. Each constituent museum unites distinguished architecture with great artworks. Looking to the future, the Guggenheim Foundation continues to forge international partnerships as it works on developing new museums in the Middle East, Latin America, and Asia (http://www.guggenheim.org/guggenheim-foundation).

This aspect implies the redefinition of the traditional centre/periphery opposition: the role each place plays depends on the flow – of information, culture, communication – and can change if the flow redirects towards another place.

> *The third important layer of the space of flows refers to the spatial organization of the dominant, managerial élites* (rather than classes) that exercise the directional functions around which such space is articulated. The theory of the space of flows starts from the implicit assumption that societies are asymmetrically organised around the dominant interests specific to each social structure. (Castells 2000: 445, my emphasis)

The form of space of the informational elite is cosmopolitism, that gives distinction in respect to the embeddedness in the local dimension. As Castells assesses, we can observe the rise of a global élite, the only one really global, as opposite to the masses, compelled in a local dimension. This aspect produces some effect in the way power is conceived in the network society.

> The fundamental form of domination in our society is based on the organisational capacity of the dominant elite that goes hand in hand with its capacity to disorganise those groups in society which, while constituting a numerical majority, see their interests partially (if ever) represented only within the framework of the fulfilment of the dominant interests. (Castells 2000: 445)

Castells proposes, on a global scale, the *iron law of oligarchy* (Michels) according to which each form of organization implies an oligarchy, which exercises power over the rest of society.

The notion of space is replaced with the notion of flows – of communication, information, capital, people. Flows are more dynamic than spaces and can reshape the space according to the direction of contents.[1]

The 'new mobilities paradigm'

Looking at the contemporary social processes, Urry proposed a 'new paradigm for social sciences' based on the notion of mobility: 'Social science presumes a "metaphysics of presence", that it is the immediate presence with others that is the basis of social existence' (Urry 2007: 47).

As Urry stressed, 'the starting point is that the analysis of mobilities transforms social science. Mobilities make it different. They are not merely to be added to static or structural analysis. They require a wholesale revision of the ways in which social phenomena have been historically examined' (Urry 2007: 44).

According to Urry:

> all social relationships should be seen as involving diverse 'connections' that are more or less 'at a distance', more or less fast, more or less intense and more or less involving physical movement. Social relations are never only fixed or located in place but are to very varying degrees constituted through 'circulating entities'. (Urry 2007: 46)

The origin of the new mobility paradigm can be found in Simmel and in the theories of complexity, postmodernism, sociology of migration, sociology of tourism, which consider the social phenomena in terms of processes. 'Social life – Urry suggests – involves continual processes of shifting between being present with others (at work, home, leisure and so on) and being distant from others. And yet when there is absence there may be an imagined presence depending upon the multiple connections between people and places' (Urry 2007: 47).

There are five different forms of mobility:

> The *corporeal* travel of people for work, leisure, family life, pleasure, migration and escape, organised in terms of contrasting time-space modalities ... The physical movement of *objects* to producers, consumers and retailers ... The *imaginative* travel effected through the images of places and peoples appearing on and moving across multiple print and visual media ... *Virtual* travel often in real time thus transcending geographical and social distance ... The *communicative* travel through person-to-person messages via messages, texts, letters, telegraph, telephone, fax and mobile. (Urry 2007: 47, original emphasis)

---

1   The notion of flows is also in Appadurai, who has stressed the five dimensions of global cultural flows: a) ethnoscapes; b) mediascapes; c) technoscapes; d) finanscapes; e) ideoscapes (Appadurai 1996).

The first two forms of mobility seem to be subjected to the old-fashioned model of physical mobility either of persons or objects; maybe they share with the last three some technological innovations. On the contrary, the three following mobilities implicitly refer to a dematerialization of space as a framework for social action. In modern societies mobility referred to movements, but it is no longer true; similarly mobility referred to physical movement (migration, nomadism) and in contemporary societies refers either to an existential condition or to the ability to manage different forms of communication at work, at home or for leisure (*The Economist* 2008).[2]

In all these cases 'the term "mobilities" refers to this broad project of establishing a "movement-driven" social science in which movement, potential movement and blocked movement are all conceptualised as constitutive of economic, social and political relations' (Urry 2007: 43). The new mobility paradigm not only refers to the increasing levels of mobility which characterize contemporary societies, but it implies the assumption that, as Simmel stressed, every social phenomenon can be understood only in its dynamic dimensions. Thus, mobility can be found in the distance that individuals have put in place in respect to their social dimensions or in the fragmentation of the identities, partially replaced by successive identification processes, as it happens in postmodern tribes (Maffesoli 1988, 1997).

According to the classical debate, that we reported at the beginning of this chapter, we could argue that the changes which have occurred in social space produce some effects in the way physical spaces are conceived. The fact that people are no longer tied to specific places expressly devoted to functions, means that there is a huge drop in demand for traditional, private, enclosed spaces such as offices or classrooms, and simultaneously a huge rise in demand for semi-public spaces that can be informally appropriated as ad-hoc workspaces (*The Economist* 2008). Rooms, walls and doors are replaced with 'open spaces' which can be reshaped by moving the social relations existing within them.

---

2   'Another big misunderstanding of previous decades was to confuse nomadism with migration or travel [...]. But although nomadism and travel can coincide, they need not. Humans have always migrated and travelled, without necessarily living nomadic lives. The nomadism now emerging is different from, and involves much more than, merely making journeys. A modern nomad is as likely to be a teenager in Oslo, Tokyo or suburban America as a jet-setting chief executive. He or she may never have left his or her city, stepped into an aeroplane or changed address. Indeed, how far he moves is completely irrelevant. Even if an urban nomad confines himself to a small perimeter, he nonetheless has a new and surprisingly different relationship to time, to place and to other people. "Permanent connectivity, not motion, is the critical thing", says Manuel Castells' (*The Economist*, 2008).

The redefinition of social action: the space of the local, the space of culture

Space, boundaries and limits

The changes, to which we referred – due to globalization, to migratory processes, communication and mass tourism – and as a consequence, the dematerialization of space, imply a redefinition of social action.

These phenomena drive individuals to cope with an unlimited notion of space. In this case, the space involves the notion of boundary, as limited space, created to define and control reality. The Nation-State is the historical product of this need of control, conducted through the framing of the space.

We can refer to the *boundary* either in a geographical (border) or in an existential sense, looking at the extension of human potentiality. The *limit* is the existential equivalent of the boundary: if the boundary is the physical space for interaction processes between the individuals, the limit is the range of human potentialities. *Boundary* is directly proportional to the *limit*: the more the *boundaries* are widened the more the sense of *limit* arises.

Early in the twentieth century the Italian author Guglielmo Ferrero suggested that the problem of modernity was to have put mankind in front of the 'unlimited' (Ferrero 1913); in his work Durkheim talked about the 'infinity evil', as a characteristic of the modern individual, moved by the will to overcome every limit, which leads to *anomy*.

The *boundary/limit* defines the framework according to which the individual can verify the consequences of his actions and work out expectations towards future actions. Technological development of mass media is enhancing human potentiality and making it possible to interact outside the circle one belongs to.

The concept of *disembedding*, proposed by Giddens, is useful to describe the new existential condition. 'By disembedding I mean the "lifting out" of social relations from local contexts of interaction and their restructuring across indefinite spans of time-space' (Giddens 1990: 21). On the one hand, this condition frees social action from the pressure of behavioural models; on the other, it means that the subject has to cope with an existential condition characterized by increasing levels of uncertainty. This uncertainty involves the construction of identity, the sense of belonging and recognition.

So the questions are: is it possible to live without boundaries? Are new boundaries emerging? The answers are difficult and ambivalent too.

New boundaries: the space of the local, the space of culture

Maybe the answer to these questions depends on individual perception of phenomena and we have to consider two main issues:

1. the emerging role of the local spaces where the local is opposite to the global; and

2. as a consequence, the new role of culture.

Different phenomena – cultural issues, religion, nationalism, fundamentalism, the claim for recognition from ethnic minorities, multiculturalism – express a process of 'culturalization', which sometimes hits extreme levels.

Globalization seems to be divided in two different dynamics: on the one hand, it seems to knock down all boundaries; on the other, it seems to rebuild other boundaries, in order not to lose control over reality. *Global* and *local* (*glocal*) are poles apart of a process of redefinition of the space, divided between limited and unlimited notions of space (Robertson 1995). Politics, consumptions and leisure are some of the examples of the strength of the local. We can also refer to the notion of *reembedding*, proposed by Giddens. As he says:

> I want to complement the notion of disembedding with one of reembedding. By this I mean the reappropriation or recasting of disembedded social relations so as to pin them down (however partially or transitorily) to local condition of time and place [...] My overall thesis will be that all disembedding mechanisms interact with reembedded contexts of action, which may act either to support or to undermine them. (Giddens 1990: 79–80)

Thus physical space is replaced by the symbolic space of culture. The notion of culture has a strong spatial meaning: it is the framework for social processes. *Culture is an instrument to reduce social complexity*, and this is the reason why it may represent a form of social exclusion.

This closed and static notion of culture implies the necessity to look at culture in terms of *cultures*, defined by the physical space. This notion of culture, which can be defined as *strong*, represents a safety net that meets a request for certainty and gives a sense of belonging to a group or to a community. The space of culture and the space of the local represent the attempt to restore some spatial forms into contemporary social phenomena. Fundamentalism, multiculturalism (as a quest for cultural identities), religion, even localism in its traditional and cultural meaning are only some examples of this tendency to re-establish boundaries.

Final remarks

The analysis of contemporary social processes is quite difficult because of the impossibility to define a univocal interpretative criterion. Hence the proposal of a new paradigm for social sciences arises.

On the one hand, all the forms of mobility – related to globalization, technology, mass media, migratory processes – seem to lead to a 'dematerialization' of space; on the other, new symbolic spatial forms are emerging such as fundamentalism, multiculturalism, and so on. The fear for uncertainty seems to be the main interpretative criterion for such a phenomenon that, otherwise, could be defined as 'regressive' if compared to other contemporary social processes.

Opposite needs face each other to build boundaries and to remove them. As Hannerz assesses, there are two aspects that seem to affect the rules of cultural organization, making them different in respect to the past: 'the mobility of human beings themselves, and the mobility of meanings and meaningful forms through the media' (Hannerz 1996: 19).

An interesting relationship between the 'dematerialization' of space and the retrieval of the space of culture is now emerging. The idea of an organic relationship between a population, a territory, a form of political organization, has been criticized as it obliges us to look at 'culture' in terms of organized packages of meanings and meaningful forms which we refer to as 'cultures' in the plural.

All the forms of contemporary mobility redefine the relationship between individuals and culture. As Hannerz assesses 'As people move with their meanings, and as meanings find ways of travelling even when people stay put, territories cannot really contain cultures' (Hannerz 1996: 8). From this point of view the opposition between global and local has been overcome by the overlapping of different levels: what is global could be local and vice versa.

The tendency to propose analyses based on such an opposition has been successful for a long time, spreading throughout the world. Unfortunately, it constitutes the main obstacle to the comprehension of contemporary social phenomena.

References

Appadurai, A. 1996. *Modernity at Large: Cultural Dimensions of Globalization*. Minneapolis-London: University of Minnesota Press.

Augé, M. 1995. *Non-places. Introduction to an Anthropology of Surmodernity*. London: Verso.

Bauman, Z. 1998. *Globalization. The Human Consequences*. Cambridge: Polity Press.

Castells, M. 2000. *The Rise of the Network Society*. London: Blackwell Publishing.

Durkheim, E. 1984. *The Division of Labour in Society*. London: Macmillan.

Ferrero, G. 1913. *Fra i due mondi*. Milano: Treves.

Frisby, D. 1992. *Simmel and Since: Essays on Georg Simmel's Social Theory*. London: Routledge.

Giddens, A. 1990. *The Consequences of Modernity*. Cambridge: Polity Press.

Goffman, E. 1959. *The Presentation of Self in Everyday Life*. Garden City, NY: Doubleday.

Hannerz, U. 1996. *Transnational Connections. Culture, People, Place*. London: Routledge.

Maffesoli, M. 1988. *Le Temps des tribus, le déclin de l'individualisme dans les sociétés de masse*. Paris: Méridiens Klincksieck.

Maffesoli, M. 1997. *Du Nomadisme, vagabondages initiatiques*. Paris: Libraire, Général Française.

Meyrowitz, J. 1985. *No Sense of Place. The Impact of Electronic Media on Social Behaviour*. Oxford: Oxford University Press.

Pacelli, D. and Marchetti M.C. 2007. *Tempo, spazio e società. La ridefinizione dell'esperienza collettiva*. Milano: FrancoAngeli.

Rifkin, J. 2000. *The Age of Access. How the Shift from Ownership to Access is Transforming Capitalism*. New York: Jeremy P. Tarcher/Putnam.

Robertson, R. 1995. Glocalization: time-space and homogeneity-heterogeneity, in *Global Modernities*, edited by M. Featherstone et al. London: Sage Publications, 25–44.

Rossi, E. 2006. *Le forme dello spazio nella tarda modernità*. Milano: FrancoAngeli.

Simmel, G. 1950. *The Sociology of Georg Simmel*, edited by K.H. Wolff. Glencoe: The Free Press.

The Economist. 2008. *Special Report on Mobility*. [online, 10 April] Available at: https://www.economist.com/surveys/displaystory.cfm?story_id=10950394 [accessed: 28 February 2009].

Urry, J. 2000. *Sociology beyond Societies: Mobilities for the Twenty First Century*. London: Routledge.

Urry, J. 2007. *Mobilities*. Cambridge: Polity Press.

Chapter 2

Mobility and the Notion of
Attainable Reach

Kjell Engelbrekt

This chapter in general terms explores the renewed interest in movement and sociopolitical transformations of space in recent social science scholarship, and British sociologist John Urry's impressive *Mobilities* (2007) in particular. Urry's *Mobilities* has the hallmarks of a lasting contribution to contemporary social science, and deserves serious treatment for its bold and comprehensive approach to a series of related issues. The argument developed below is one that, above all, readily accepts the conceptual and theoretical challenge posed by the sociology of mobility and associated transformations of communications and space. Though the chapter critically examines some pivotal aspects of Urry's argument, it primarily tries to render the terminology of sociopolitical transformations more precise by following up on Urry's desire to relax the conceptual boundary between mobility and immobility, as well as that between presence and absence.

Whereas several new sociologies of movement, communication and space are in the process of 'recoding' problematics that in traditional social theorizing rely on an ontology of proximity, boundaries, immobility or permanence (Boden and Molotch 1994, Torpey 2000, Massey 2005, Shain 2007, Castles and Miller 2008), Urry has inserted a premise of mobility and impermanence of social existence that radically reverses the 'burden of proof' in the epistemological sense. As he draws on the classical writings of Georg Simmel as well as those of Robert E. Park and the so-called Chicago School, Urry outlines what he refers to as a '"movement-driven" social science in which movement, potential movement and blocked movement are all conceptualized as constitutive of economic, social and political relations' (Urry 2007: 43). Another important contribution inherent to the sociology of mobility is that of reducing the analytical distance between 'co-presence' and 'connected co-presence', the latter referring to means of virtual communication and travel brought about by modern information technologies.

In this last respect the new sociology of mobility has considerable potential, as it could easily appropriate additional elements that would allow for a theoretical and analytical extension in the said direction. A straightforward possibility would be to introduce a larger portion of the terminology of 'life-world' phenomenology, based on the rich German tradition of social theory and philosophy. One of the conceptual building blocks for such an exercise is precisely Simmel's concept of 'co-presence', borrowed and put to good use by Urry himself but, arguably, in certain respects further elaborated by Alfred Schutz and Thomas Luckmann in *Structures of the Life-World* (1973). Similarly, a conceptual pair that appears directly relevant for the new sociology of mobility is 'attainable reach' and 'restorable reach', both of which denote the distance to 'the starting point for my orientation in space' (Schutz and Luckmann 1973: 46). The chapter ends with an attempt to render certain phenomenological concepts compatible with the new sociology of mobility, so as to make them useful for empirical research or for construction of a more comprehensive theory.

Connections

Modernity, according to a formulation by Anthony Giddens, means that 'larger and larger numbers of people live in circumstances in which disembedded institutions, linking social practices with globalized relations, organize major aspects of day-to-day life' (Giddens 1990: 79). Yet in the same text Giddens goes on to supplement this definition with the notion of 're-embedding', a process equivalent to 'the reappropriation or recasting of disembedded social relations so as to pin them down (however partially or transitorily) to local conditions of time and place' (Giddens 1990: 79–80). In turn, he links the notion of reembedding social relations to other key features of modernity, namely 'trust in abstract systems' and 'faceless commitments'. Finally, Giddens writes of 'access points' at which a connection is made between 'representatives of abstract systems' and 'lay individuals' (Giddens 1990: 88).

The early 1990s were a time of manifold attempts at reconsidering the condition of modernity, with Giddens expressing one of the most eloquent, synthesizing arguments. It was a time in which the virtual explosion of literature devoted to the problem of globalization had not yet occurred, but when terms like interdependencies, networks and connections were increasingly used as theoretical building blocks in the social sciences. The forward-looking work of Paul Virilio on speed and circulation was underway (Virilio 1986, 1991), and Manuel Castells had begun writing his ambitious three-volume contribution on the communications revolution and the associated network society (Castells 1996, 2001, 2004). Meanwhile, among a number of observers, there were rapidly growing concerns about corporate-led capitalism and its medium- to long-term implications on state institutions and social life (Hirst and Thompson 1996).

In *Mobilities* Urry has in a fruitful way connected to some of these earlier debates but also steered free of other, less helpful ones. Despite his synthesizing ambitions, he has managed to concentrate rather effectively on the task of theorizing physical travel and virtual communication. The result is a focused but at the same time encompassing sociology of mobility, that aspires to extend its range beyond the limitations imposed by 'nomadic' and 'sedentary' ontologies and frameworks. This means that Urry emerges as equally wary of positions that rely on a premise saying that society faces large-scale unsettling of social life resulting from enhanced mobility in the direction of (ever-)increasing fluidity, as he is of the notion that a return to a pre-globalization situation of limited movement is feasible. The sedentary position, in its most sophisticated forms often directly or indirectly derived from Martin Heidegger's social philosophy, is thus an implicit object of Urry's criticism. The nomadic position, for its part, can be associated with passages in the work of Zygmunt Bauman (1998) and perhaps that of Majid Tehranian. At one point Urry explicitly criticizes nomadism for its tendency toward a 'fetishism of movement' (Urry 2007: 186).

In the 1990 *Consequences of Modernity* Giddens also talked of 'faceless commitments' as 'trust relations which are sustained by or expressed in social connections established in circumstances of copresence' (Giddens 1990: 80). Urry largely avoids the complex notion of trust but otherwise agrees with Giddens, Simmel, and further with Deidre Boden and Harvey Molotch (1994) on the idea that co-present interaction remains the fundamental mode of human intercourse and socialization (Urry 2007: 24). He also acknowledges, despite his critique of sedentary ontologies, the usefulness of Heidegger's distinction between objects that are 'ready-to-hand' as opposed to objects 'present-at-hand' (Heidegger 1962, cited in Urry 2007: 45). In this particular context he quotes Keller Fox, who suggests that some of the ready-to-hand tools of modern communication are indeed likely to help recreate the more human, natural patterns of communication that prevailed in pre-modern times, with 'gossip' as a key feature (Fox 2001, cited in Urry 2007: 175).

Social life is being profoundly affected, Urry explains, through the daily use of 'mundane virtual objects' like cell phones, computers, electronic organizers, and the like (Urry 2007: 162). Urry does not go as far as Fox in expressing appreciation for the ability of modern communication technologies to restore 'natural', pre-modern patterns of human interaction, but points out that communication technology often facilitates the organizing of co-present meetings (Urry 2007: 172). He further insists that co-present and distant connections today increasingly intermingle (Urry 2007: 177). Moreover, the time horizon of social relations is said to be altering as a result of recent technological innovations. Just like clocks and watches once created the modern notion of punctuality, Urry refers to research illustrating that the cell phone, for instance, is re-establishing 'fluid and negotiated time' (Urry 2007: 172). Through e-mail and text messages, there can also be a sense of connectedness even though the persons participating in the dialogue live many time zones apart and may therefore have limited opportunities for real-time interaction.

One of the most ambitious theoretical claims made in *Mobilities* is that Urry wants us all to move away from what he sees as unfruitful types of dichotomies that obscure the emerging patterns of societal change. He lists five such relevant dichotomies: real versus unreal, face-to-face versus life on the screen, immobile versus mobile, community versus virtual, and presence versus absence (Urry 2007: 180–81). A straightforward objection to this sweeping assertion in favour of relaxing dichotomies might be that Urry himself establishes and utilizes the immobile/mobile distinction as the fundamental premise of his book, and thereby can be accused of reasserting this particular dichotomy all over again. But in all fairness, this objection would ignore the circumstance that most social science theorizing indeed appears to suffer from an 'immobility bias', along with most Western European philosophy. So this criticism may not only be oversimplifying Urry's argument, but seriously be playing down the analytical potential of the mobilities approach.

In order to better assess the approach on its potential for opening up thinking space and generating empirical studies, we should therefore now turn to the more detailed framework of analysis that Urry is providing us with in the 2007 volume. In this regard a seemingly potent typology is available in what the author himself refers to as the five 'interdependent mobilities that produce social life organized across distance and which form (and re-form) its contours' (Urry 2007: 47). The five interdependent mobilities described by Urry are 'corporeal travel', 'the physical movement of objects', 'the imaginative travel effected through images [...] and people appearing [in] media', 'virtual travel [...] transcending geographical and social distance', and 'communicative travel through person-to-person messages' (Urry 2007: 47). At least ostensibly, the typology is poised to transcend the five dichotomies listed above, and the bulk of the book takes us through how this can be accomplished. As Urry summarizes his conclusions with regard to the interdependent mobilities in the last chapter, he is able to convincingly demonstrate both the theoretical distinctiveness and the analytical pertinence of his approach (Urry 2007: 272–6).

Communion

Beyond his ambition to relax the boundary between the five dichotomies just mentioned, Urry tries to get beyond nomadic and sedentary positions by offering a distinction between three different senses of community. First, there is community in the straightforward topographical sense. This corresponds to 'settlement based upon close geographical propinquity, but where there is no implication of the quality of the social relationships found in such settlements of intense co-presence' (Urry 2007: 163). Second, there is 'the sense of community as any local social system in which there [is] a localized, relatively bounded set of systemic interrelationships of social groups and local institutions', as the author puts it (Urry 2007: 163). Third, there is communion, Urry says, or more precisely 'human association characterized by close personal ties, belongingness and emotional warmth between its members. [...] This is community as "affect"' (Urry 2007: 163).

Urry has evidently borrowed the notion of 'communion' from Colin Bell and Howard Newby (1976) to describe 'the more distant and intermittent connections that hold social life together', an aspect that he convincingly argues is otherwise weakly developed in much contemporary sociological thought (Urry 2004: 26). All mobilities require economic resources, he adds, emphasizing the trend of accelerating bifurcation of society into different socioeconomic groups. Not unlike the term 'digital divide', illuminating a deepening gulf between the haves and have-nots of communication technology, Urry talks of 'hot and cold' economic zones in relation to the technological level of relevant information infrastructures (Urry 2007: 191–2). Since this distinction does not allow for fine-tuned analysis, though, Urry goes on to introduce a list of eight different elements of network capital. Similar to Robert Putnam's more famous, parallel concept of social capital, the notion of network capital takes the socioeconomic capabilities of the relevant groups into account (Urry 2007: 196–8).

Urry is careful to distance himself from Robert Putnam's diagnosis of contemporary Western society as one in which social capital and civil society institutions are stagnating or even eroding. But he nevertheless seems sympathetic to pessimistic accounts made by other observers, paying attention to an overall correlation between the weakening of the state and its welfare system on the one hand, and civil society and patterns of social relations on the other. In citing Hirst and Thompson (1996), moreover, Urry goes as far as to say that international movement hollows out 'civil society and its organizing power over both the life-chances and the life-styles of its "members"' (Urry 2007: 189).

Concerning the technology of communication and travel, though, there is a less pessimistic outlook. At times Urry's arguments could appear somewhat reminiscent of Virilio's phenomenologically grounded depiction of social alienation through technology, as well as of the latter's peculiar admixture of humanism and materialism (Virilio 1986, 1997). But in important respects his view differs from that of Virilio, who primarily laments the replacement of ocular observations with technologically mediated perceptions and representations (Virilio 1991: 111). In fact, Urry can be seen as both embracing and problematizing the theoretical challenges involved in probing the impact of technologies of mobility on social life. Given his nuanced approach, a more pertinent question to ask is whether Urry's three-fold typology represents an adequate solution or response to this challenge. For my own part, I believe three significant criticisms can be raised against Urry's typology and its integration of the concept of 'communion', in that the former seems to include contradictory elements.

The first criticism would be that the first two categories of the typology more or less rely on classical, sedentary conceptions of community. The use of terminological elements such as 'settlements', 'geographical propinquity', as well as 'local social system' and 'localized [...] systemic interrelationships' seem to evoke precisely the sedentary-oriented half of the dichotomies Urry sets out to dismantle, in that they are characteristically associated with real, face-to-face, immobile, community-based and co-present social relations. If Urry wants to relax the boundaries between the said dichotomies, it would seem necessary to avoid reinserting them 'through the backdoor' by virtue of analytically oriented typologies and frameworks.

A second criticism relates to the tension that exists between 'community as affect', the equivalent of which Bell and Newman refer to as 'communion', and Urry's own observation that 'weak ties' are the increasingly dominant mode of social interaction, especially among the young. It is not entirely clear whether his position in this respect is a nuanced one or whether it contains contradictory elements and eventually becomes inconsistent. Toward the end of his book Urry details the ongoing expansion and growing significance of different forms of network capital, of 'personalized' network patterns, of the perception that 'whom you know' is widely more important than 'what you know', and of the 'society of the schedule' (Urry 2007: 229, 245). These observations may all be valid, but they underline a trend toward 'weak ties', that is, toward social relations that are lacking in some aspects of 'authentic' community spirit. While Urry frequently and appreciatively quotes Barry Wellman's assertion that social capital is being built through and within virtual networks (Wellman 2001), he also rejects or tries to undermine claims by Putnam or Richard Sennett that enduring friendships and relations of trust are declining due to the flexibility requirements imposed by that same technology (Sennett 1998).

Finally, a third and more constructive critical comment would be that an elaborate sociology of mobility and space needs to take into account how people experience and perceive a sense of connection or belonging. Methodologically speaking, part of Urry's theorizing resonates closely with the macro-sociological gaze of Herbert Spencer and Émile Durkheim, whereas other parts seem inspired by Simmel and German phenomenologists as well as by Chicago School 'ecological' sociology. The affinities with the Chicago School are something that the new sociology of mobility has in common with urban sociology at large, and which in this context can be illustrated by recalling Roderick McKenzie's definition of mobility as 'a measure both of expansion and [social] metabolism, susceptible to precise quantitative formulation, so that it may be regarded almost literally as the pulse of the community' (McKenzie 1967: 61). Clearly lacking in this definition is the micro-sociological perspective that *Verstehen* approaches may offer. In the remainder of the chapter, I will therefore argue that Urry's approach can benefit from a methodologically closer relationship to Simmel and, more generally, German phenomenology.

Life-world phenomenology: potential reach

One such *Verstehen* approach, deeply anchored in German social philosophy, is available in Alfred Schutz's and Thomas Luckmann's 1973 *Structures of the Life-World*. Schutz and Luckmann regard the human mind as predisposed toward maintaining the status quo and reapplication of solutions that already have been tried and tested. Vital to Schutz and Luckmann's theory of the structure of the life-world is thus 'the natural attitude', described as a pragmatic orientation aimed at 'mastering' the situation of the day. As soon as mastering is seen as possible and accomplishable without replacements in the gallery of so-called 'sedimented types', the 'and-so-forth idealization' (as described by Edmund Husserl, the founder of phenomenological philosophy) meaning that tried-out solutions coupled with a *ceteris paribus* assumption will be the most attractive option 'until further notice'.

Nonetheless, a pragmatic orientation aimed at mastering the situation does not necessarily imply conservatism or an obsession with the present. Almost to the contrary, Schutz and Luckmann posit that our knowledge 'makes possible the orientation of the flux of experience toward the future'. Our constant search for 'typicality' in everyday situations is strongly motivated by this forward-looking orientation, since 'types' help us guide our future action. Indeed, they argue that 'the typical and typically repeatable aspects of experience and of action are of interest' (Schutz and Luckmann 1973: 241). So although it is true that most experiences fall within the established 'types' or routine situations, the world repeatedly imposes novel elements that render situations 'problematic' (Schutz and Luckmann 1973: 122–6). When existing types thus are challenged, the relevant 'sedimented' or, in another phrase, 'past lived' experiences are adjusted to accommodate new information to which we are exposed.

The logic of the life-world is not merely related to experience of time, as in sedimented types, but to experience of space as well. 'The place in which I find myself, my actual "here"', Schutz and Luckmann write at one point, 'is the starting point for my orientation in space' (Schutz and Luckmann 1973: 36). There can in fact be no life-world without such a starting point or centre in spatial and temporal terms, as once theorized by George Mead (1932: 124ff). This is the world of *actual reach*, the world of which I have intimate and immediate knowledge, and upon which I can act directly. A second portion of the life-world, that of which I similarly have unmediated experience but which presently is beyond direct access, is referred to as the world within *restorable reach* (Schutz and Luckmann 1973: 36–8).

If actual and restorable reach are fairly straightforward categories, the concept of *attainable reach* denotes a more complex arrangement. In principle, for an individual there is no absolute obstacle to my attaining access to any part of the world (for the sake of simplicity here understood as the planet earth). But there are considerable variations of attainability in terms of subjective degrees of probability as well as ability. The arrangement into degrees or sublevels of potentiality of reach depends, according to Schutz and Luckmann, on the perceived prospect of attainability. Since our expectations are based on our stock of knowledge, consisting of sedimented experiences, the perceived spatial, temporal and social distance will inevitably reflect the past and present states of anticipation in this respect (Schutz and Luckmann 1973: 39–40).

At the same time, a steadily expanding scope of unmediated interaction with other social subjects and spaces will consequently harbor the potential of precipitating a shift from a low to a high level of attainability. Since the zones of actual and restorable reach are logically bound up with a wide variety of immediate interaction, a rapid growth in such contacts would seem to entail considerable implications for people located at the centre of this social process. It would have the capacity of affecting differentiations in terms of intimacy and anonymity, strangeness and familiarity, along with social proximity and distance (Schutz and Luckmann 1973: 41). The orbit of actual and restorable reach can in fact only be extended through unmediated interaction, and the same goes for levels of attainability.

As could be expected, the categories of Schutz and Luckmann have already been applied to sociology of travel and sociology of communication. At a conceptual level Wellman and Haythornthwaite used phenomenological reasoning along the lines of Schutz and Luckmann to discuss a variety of aspects in the activities of online networks (Wellman and Haythornthwaite 2002). Bakardjieva, meanwhile, applied the framework more comprehensively in a qualitative study of home users of the Internet. Bakardjieva could show, among other things, that the boundaries between private and public become subject of a subtle negotiation for every individual user (Bakardjieva 2005: 180–81). She also found that virtual groups for the most part do not displace face-to-face communities but fill a gap of presently inaccessible social relationships, or such relationships that for one reason or another cannot satisfy the needs of its members (Bakardjieva 2005: 178–9).

Neither Wellman and Haythornthwaite nor Bakardjieva, however, systematically explore the notion of co-presence in their respective studies. By virtue of his broader theoretical ambitions, Urry is well placed to theorize economically and technologically induced sociopolitical transformations in relation to phenomenological terms such as co-presence and potential reach. From the vantage point of the mobilities paradigm, Urry already speaks of 'connected co-presence' in the sense that 'others are there but at a distance' (Urry 2007: 212). But more importantly, he goes onto qualify the latter notion by suggesting that communication tools facilitate this co-presence under conditions of 'intermittent meetings' or by emphasizing the 'intensive scale' of human interaction (Urry 2007: 216). Toward the end of the book, this argument is further elucidated as Urry discusses the feasibility of thick intermittent co-presence that 'involves rich, multi-layered and dense conversations', that is, by human interaction of a certain quality (Urry 2007: 236).

With reference to such thick, rich, multi-layered and dense conversations, Urry thus reconciles what in other parts of the literature often remain starkly opposing views. While privileging the two conventional, 'sedentary-oriented' senses of community over communion, he explicitly accepts the notion that network (and social) capital indeed accrue in online collectivities. Such collectivities, Urry writes, are like 'new fractal social spaces, as each realm folds over, under, through and beyond each other in striking new social topologies. These are oscillatory, flickering, both here-and-there, both inside and outside, rather like a Mobius strip' (Urry 2007: 181). The argument is that instances of connected co-presence ultimately rely on some form of 'intense' social interaction and intermittent face-to-face meetings, regardless of which precedes the other. In the end, therefore, Urry effectively acknowledges a human need to engage in rich, multi-layered and dense co-presence, a co-presence the characteristics of which are more easily sustained through modern mobilities; on the other hand, multi-layered and dense co-presence rarely arises out of 'weak ties' alone (Urry 2007: 212).

Attainability: whom to reach and why?

In lieu of a conclusion, I would like to make a couple of remarks aimed at rendering compatible Urry's imaginative sociology of mobility with life-world phenomenology, and so try to enhance the analytical potential of both.

Out of the five interdependent mobilities described by Urry, 'corporeal travel of persons' and 'physical travel of objects' would seem to lend themselves to being subjected to old-fashioned physical barriers and administrative restrictions of various kinds. The ease with which many people, goods and financial assets today can be made to 'travel' is unprecedented, but there are still means with which governments and international organizations can curb such flows. The same is clearly not true for the other three interdependent mobilities, namely 'imaginative travel through images', 'virtual travel [...] transcending geographical and social distance' or 'communicative travel through person-to-person messages' (Urry 2007: 47). The latter three share some distinctive qualities in comparison with the previous two, qualities bolstered by economic and technological development in the past two decades or so.

The terms 'movement', 'potential movement' and 'blocked movement' may be helpful in disentangling those particular qualities, and also bringing them into closer correspondence with life-world phenomenology (Urry 2007: 43). Whereas 'movement' and 'blocked movement' tend to apply to objects of the physical world, including individuals (Cunningham and Heyman 2004), as well as to reflect a sedentary sociopolitical logic of rationality, the latter three interdependent mobilities are directly intertwined with the 'social world' and with 'potential movement'. At this point it appears appropriate to reinsert a term from the parlance of Schutz and Luckmann, namely that of 'the world of potential reach'.

As mentioned above, the world of potential reach may be further divided into two distinct subcategories, the worlds of restorable and attainable reach, respectively. The first subcategory, the world of restorable reach, mainly depends on the human capacity to remember and imaginatively 'revisit' a space to which he or she has had direct and unmediated access sometime in the past. The notion thus acknowledges the intricate workings of the human intellect as well as the rich and multi-layered powers of perception derived from having physically experienced a location, or a fellow human being, in co-presence. Already more than 35 years ago, Schutz and Luckmann insisted that the orbit of actual and restorable reach can only be extended through such unmediated interaction. As far as I can see, this assertion resonates closely with Urry's notion of 'intermittent connections' or co-present encounters in order to achieve 'thick' human interaction.

By contrast, the world of attainable reach is a more complicated notion, but appears even more relevant to imaginative, virtual and communicative travel. The world of attainable reach would seem to correspond quite well to Urry's 'connected co-presence' and the idea that 'others are there but at a distance' (Urry 2007: 212). All of the latter three types of travel, one could also argue, may expand the horizon of the social world many times over by force of the formidable economic and technological developments of recent years. To take a simple example, physical boundaries and legal restrictions can be applied and bar access to the newest computer software, and thereby limit the opportunities to play sophisticated video games on expensive computer platforms for users in many developing countries. On the other hand, that same computer user can still find a way to (illicitly) download the latest blockbuster movie fresh out of the Hollywood industry.

More importantly, however, imaginative, virtual and communicative transcend geographic and social distance to a degree not experienced in the past. This is where Urry's 'mundane virtual objects' need to be brought into the picture. Let us take another example. For instance, the possession of a Chinese or Bangladeshi passport does certainly not buy a citizen of those countries a ticket and a visa to Paris, France. By the same token, though, for millions of Chinese and Bangladeshi citizens attainability may today be there on a level never before experienced. Supposedly in today's China or Bangladesh, learning French as a school subject and working hard at creating good career opportunities after college may very well place a coveted 'space of attraction', such as the French capital, within a reasonable distance (Urry 2007: 253). Albeit the goal is distant, objectively speaking, in the social world it can very well be within 'attainable reach'.

In the meantime, there is the Internet and a host of evolving software applications that partially bridge geographic and social distance to an unprecedented degree. In his brief discussion of the history of street walking and consumption in urban public spaces, Urry describes the nineteenth-century Parisian *flâneur* or *flâneuse* as the forerunners of contemporary tourists (Urry 2007: 69). Indeed, a virtual tourist couple of the early twenty-first century may now stroll around the World Wide Web in ways not wholly different from those of a French metropolitan hero in the past, gazing at Haussmann's boulevards via the Google Earth (Panoramic or Street View) software or at individual buildings by way of the thousands of tourist images available at www.virtualtourist.com and other websites where individual photos can be posted.

This is of course not to suggest that 'connected co-presence' would be a satisfying substitute of co-presence in the qualitative, phenomenological sense. Few of today's virtual *flâneurs* are supposedly inclined, after engaging in the latter type of excursion, to turn to a fellow *flâneuse* – as Rick (Humphrey Bogart) to Ilsa (Ingrid Bergman) in *Casablanca* – and conclude: 'We'll always have Paris'. The rich, multi-layered and dense experience that we seem to be gleaning in Ilsa's eyes is in other words unlikely to be derived solely from the world of attainable reach, in Schutz's and Luckmann's terms. For Rick is alluding to more than one interdependent mobility but more importantly to the world of restorable reach, as he completes the thought: 'We didn't have it before ... we'd ... we'd lost it until you came to Casablanca. We got it back last night'.

References

Bakardjieva, M. 2005. *Internet Society: The Internet in Everyday Life*. London: Sage.

Bauman, Z. 1998. *Globalization: The Human Consequences*. London: Polity Press.

Bell, C. and Newman, H. 1976. Communion, communalism, class and community action: the sources of the new urban politics, in *Social Areas in Cities*, edited by C. Bell et al. Volume 2. Chichester: Wiley.

Boden, D. and Molotch, H.L. 1994. The compulsion to proximity, in *Nowhere. Space, Time and Modernity*, edited by R. Friedland and D. Boden. Berkeley, CA: University of California Press.

Castells, M. 1996. *The Rise of the Network Society*. Maiden: Blackwell.

Castells, M. 2001. *The Internet Galaxy: Reflections on the Internet, Business, and Society*. Oxford: Oxford University Press.

Castells, M. 2004. *The Power of Identity. Information Age: Economy, Society, and Culture*. Volume II. London: Blackwell.

Castles, S. and Miller, M.J. 2008. *The Age of Migration: International Population Movements in the Modern World*. Fourth edition. New York: Guilford Press.

Cunningham, H. and Heyman, J. McC. 2004. Introduction: mobilities and enclosures at borders. *Identities: Global Studies in Culture and Power*, special issue, 11(3), 289–302.

Giddens, A. 1990. *The Consequences of Modernity*. Cambridge: Polity Press.

Hirst, P. and Thompson, G. 1996. *Globalization in Question: The International Economy and the Possibilities of Governance*. Cambridge: Polity Press.

Massey, D. 2005. *For Space*. London and Thousand Oaks, CA: Sage.

McKenzie, R.D. 1967 [1925]. The ecological approach to the study of human community, in *The City*, edited by R.E. Park and E. Burgess. Fifth impression. Chicago: University of Chicago Press.

Mead, G. 1932. *Philosophy of the Present*. Chicago: Aspen Court.

Park, R.E., Burgess, E. and McKenzie, R. 1967 [1925]. *The City*. Fifth impression. Chicago: University of Chicago Press.

Putnam, R. 1993. *Making Democracy Work*. Princeton: Princeton University Press.

Schutz, A. and Luckmann, T. 1973. *Structures of the Life-World*. Evanston: Northwest University Press.

Sennett, R. 1998. *The Corrosion of Character. Personal Consequences of Work in the New Capitalism*. New York: W.W. Norton and Company.

Shain, Y. 2007. *Kinship and Diaspora in International Affairs*. Ann Arbor: University of Michigan Press.

Torpey, J. 2000. *The Invention of the Passport: Surveillance, Citizenship, and the State*. Cambridge: Cambridge University Press.

Virilio, P. 1986. *Speed and Politics: An Essay on Dromology*. New York: Semiotext(e).

Virilio, P. 1991. *The Lost Dimension*. New York: Semiotext(e).

Virilio, P. 1997. *Open Sky*. London: Verso.

Urry, J. 2007. *Mobilities*. Cambridge: Polity Press.

Wellman, B. 2001. Physical space and cyber place: the rise of networked individualism. *International Journal of Urban and Regional Research*, 25(2), 227–52.

Wellman, B. and Haythornthwaite, C. (eds) 2002. *The Internet in Everyday Life*. Oxford: Blackwell Publishing.

Chapter 3

The Harvest of Dionysus:
Mobility/Proximity, Indigenous Migrants
and Relational Machines

Carmelo Buscema

Introduction

The purpose of this chapter is to suggest an interpretation and, consequently, a definition of the pivotal concepts of 'mobility' and 'proximity' on the basis of Karl Marx's concept of 'socialization of work', Carl Schmitt's conception of the elementary dialectic between 'land and sea', and Michel Foucault's theorization of 'bio-power'. The general aim inherent in this operation is to contribute to the delineation of a set of epistemological and heuristic categories able to interpret the most relevant social, political, and economic phenomena distinguishing our epoch. In particular, herein we want to highlight the ongoing emergence of a new dimension of spatiality, characterized – from a sociological point of view – by a new relational status and by novel criteria and dynamics of separation and exclusion, mobility and resistance, inclusion and articulation – again, at the social, political, and economic levels. The importance of the three considered authors for such a perspective consists of their particular conception of space (and time) not merely in bodily terms, but as a complex relational sphere intimately transcended by opposing social, political, and economic forces, instances, and subjectivities: namely, those aimed at establishing order and mechanisms of accumulation, and those fighting expropriation and alienation through ever changing forms of resistance and escape. Nevertheless – as will be argued – the way we consider the mobility/proximity dynamic articulation is transversal and intrinsic to that opposition. In fact, we cannot simply think of 'mobility' as a synonym for 'movement' or, even less, for 'freedom of movement'. Rather, it is a social phenomenon positively articulated and historically shaped by those struggles.

It is crucial, in our opinion, to analyse the synchronic and diachronic dialectic between the instances of 'mobility' and 'proximity', since we assume they are the essential dynamics of capitalism that tune its models and points of equilibrium both on the historical and the spatial planes. Moreover, the research of the proper combination of the processes of *mobilization of the social forces* and of *articulation of their interconnectedness* (proximity) can be easily seen as the essential vector of the historical development of the modes of creating surplus value – as will be argued. In the following pages, we will distinguish three main stages of such a dynamic combination.

In the second part of this chapter, we will offer an exemplification of the above mentioned phenomenon of transformation of the relational and spatial reality, ongoing in the contemporary world. It will be taken into account, in empirical terms, how the current arrangement of mobility and proximity is truly blurring traditional fundamental distinctions such as those between rurality and urbanity, global peripheries and centre, advanced productive sectors and 'spheres of subsistence'. For this purpose, we will describe the main results of the fieldwork activities carried out – in California, USA, and in Mexico – on the emblematic case of the Mixtecos transnational migrants from Oaxaca. It will be demonstrated how the relational status of the indigenous communities is being enriched by the initiative of their own subjectivities, now acting on the social and historical plane of a novel articulation between instances of mobility and proximity – which they actively contribute to shape and move forward through migration and social appropriation of technologies. We will conclude that the *social appropriation* of the new technological apparatuses – what are called *relational machines* (Buscema 2005a) – by indigenous migrants, represents their ultimate endowment to the *creation of the world* (Nancy 2003) and to the realization of the *multitude* (Hardt and Negri 2001, 2005). The latter concept has to be conceived as the new articulated social productive body, adequate to the globalization regime and the postindustrial mode of value creation; while relational machines represent those individual appendices potentiating human faculties and abilities, and constituting the novel fabric of the current hypertrophied sociality.

According to the Greek and Roman mythologies, Dionysus is the divinity of the natural and mysterious energies making fruits grow and human pleasures and prosperity available – even though unequally distributed throughout society. We will argue how such a divinity can represent the proper allegory of the *workforce* in the capitalistic system – both as a concept and a concrete condition of existence. Thus, the harvest of Dionysus could be thought of as the *actual potential* of emancipation and strength engendered worldwide by the quite generalized faculty of communicating, producing and sharing wealth, and co-constructing alternative senses of development.

Finally, the emphasis we put on such processes is not intended to be celebrative: it is rather motivated by the assumption that they are the new arena within which the novel political agenda is being written, where the ulterior social antagonisms are effectively maturing and – consequently – upon which the scientific analysis must be focused.

The socialization of labour

First of all, from a Marxian perspective, the continuous transformation of all social, political, and cultural conditions is an essential feature of capitalism, which needs to continuously revolutionize all social and productive relations in order to produce surplus value. Actually, for Marx, 'capital' is a dynamical set of ever-extending social relations mediated by commodities, progressively creating the *socialization of labour* or – as we suggest to interpret such a category nowadays – the *generalization of the productive process* of all sorts of values.

Specifically, the expansion of capitalism consisted of the strict and mutual articulation of ever-increasing *mobility* and ever-increasing *proximity*. The latter is the interconnection of each and every individual or producer within the global market, progressively set by capitalism; and the former represents the social ability or historical practice of being on the move, as required by the capitalistic need for continuously revolutionizing all social and productive relations and for structuring ever-changing articulations of the division of labour.

In the following quotation from the *Manifesto of the Communist Party*, Marx and Engels reveal capitalism's inextinguishable need both for making everything *mobile*, and complementarily for getting its new criteria of *proximity*, or relational articulation, positively settled and established everywhere:

> The bourgeoisie cannot exist without constantly revolutionizing the instruments of production, and thereby the relations of production, and with them the whole relations of society. [...] Constant revolutionizing of production, uninterrupted disturbance of all social conditions, everlasting uncertainty and agitation distinguish the bourgeois epoch from all earlier ones. All fixed, fast frozen relations, with their train of ancient and venerable prejudices and opinions, are swept away, all new-formed ones become antiquated before they can ossify. All that is solid melts into air, all that is holy is profaned, and man is at last compelled to face with sober senses his real condition of life and his relations with his kind. The need of a constantly expanding market for its products chases the bourgeoisie over the entire surface of the globe. It must nestle everywhere, settle everywhere, establish connections everywhere. (Marx and Engels 1848: 5–6)

According to Marx, the process of historical and spatial expansion of the capitalistic system – where mobilization of everything is oriented to its settlement and, dialectically, its settlement implies a continuous mobilization of all social elements – can be described as the socialization of labour, basically consisting of the following three steps:

- the progressive erosion or erasing of those social, political, and physical barriers within which social, economic, political, and cultural localisms of the traditional communities developed themselves;
- the progressive setting up of a world market, within which each individual tended to be connected with every other one, and all of them mutually constrained within a general relation of material and reciprocal dependence;

- in summary, the formation and affirmation of the reality of 'capital', conceived and consisting of a (ever extending) general social relation mediated by things which historically become commodities.

In particular, Marx distinguishes two different aspects of such a general process, which are at the same time two logical and historical instances. We have primitive accumulation, on the one hand, and the division of labour, on the other, which can be respectively thought of as the *pars destruens* and the *pars construens* of the overall process.

The first 'event' describes the transformation of goods and people, respectively, into commodities and workforce, both becoming in this way saleable and buyable on the market. Actually, according to the traditional relations of power, land and individuals were mutually and strictly tied within the scheme of personal dependence centred on the feudal lord and traditional values; otherwise, some terrains and natural resources were residually conceived as common or public goods, freely accessible to members of the communities. Such a process implied the violent separation of huge masses of people from their accustomed means of production and subsistence, and consequently a vast phenomenon of pauperization and vagabondage, that we can easily consider as the very first stage of capitalistic mobility.

In his analysis of this process, Marx gives also account of how those masses of people and their flows were governed, and functionally addressed: their irregular movements were not blocked, but 'artificially' constrained within guides or funnels, forcing people to look for new solutions of subsistence in the labour market of urban areas and of those places where the manufacturing plants were settled and growing. Such a degree of social unintended movement and disorder became 'mobility' – i.e. a social phenomenon, politically controlled and systematically made productive – throughout the intervention of the State, and of its 'bloody legislation'. A more detailed account of the role played by the State within this process is also offered by Polanyi's *Great Transformation* (1944) – which refers to the constitution of the market as an autonomous institution, no longer socially regulated.

In order to better understand the deep difference existing between what we here call 'mobility', on the one hand, and 'movement' or 'freedom of movement', on the other, it is useful to take into consideration the difficulties which the settlement of capitalism had to face in North America. As Wakefield (1833) pointed out, there the existence of an open *frontier* – instead of the European closed boundaries – and the consequent possibility for people to freely move through it, to push it forward, and to own land by themselves, for a long time prevented the existence of a proletariat to be exploited (at least, until the intervention of the State, as Marx noted). The social experience of 'movement' *performed by people throughout* the American 'frontier' was not comparable at all with the social experience of 'mobility' *suffered by people within* the European boundaries.

The role played by the State within this process exemplifies what – referring to the affirmation of capitalism and the socialization of labour – we can call the *pars construens*. Actually, the implementation of a new social order based on waged labour as the source for subsistence, and on the surplus accumulation through the exploitation of the cooperative work, represented the premise for the effective constitution of the 'capital' as an overall social relation mediated by commodities. Moreover, that social order represented the concrete productive articulation of capital, which took the form of an increasing *division of labour* – to be thought of as the specific result of a social cooperation mediated and positively organized by the capitalists.

The development of modern machineries – and their systematic application to the industrial productive procedure – represents the synthesis of the general process of *socialization of labour* we have herein summarized. In fact, modern machinery is the very emblem of the *socialized work* as long as we conceive it as the new depositary of the knowledge expropriated from workers, and made objective engine of production; as 'dead labour' crystallized in ever-available and well-behaved technology; as the critical point of encounter of the divided and specialized activities of each individual worker or producer of the world market. Now, machinery's subject is 'capital' as a set of relations, and its appendage-executor is no longer the skilled worker of the manufacturing system, but 'man in general', or more generic human faculties. Marx wrote in *Capital*:

> The implements of labour, in the form of machinery, necessitate the substitution of natural forces for human force, and the conscious application of science, instead of rule of thumb. In Manufacture, the organisation of the social labour-process is purely subjective; it is a combination of detail labourers; in its machinery system, Modern Industry has a productive organism that is purely objective, in which the labourer becomes a mere appendage to an already existing material condition of production. In simple co-operation, and even in that founded on division of labour, the suppression of the isolated, by the collective, workman still appears to be more or less accidental. Machinery, with a few exceptions to be mentioned later, operates only by means of associated labour, or labour in common. Hence the co-operative character of the labour-process is, in the latter case, a technical necessity dictated by the instrument of labour itself. (Marx 1887: 252)

Machinery and the big modern factories were thus the propulsive centre of a second and more extended stage of the dialectic between mobility and proximity. It was the emergence and evolution – between the second half of the nineteenth century and almost the entire twentieth century – of a productive system structured all around the effect of the industrial systematic application of a new technological turn which contributed to change the very nature of labour. Then, the big automatic machines gained the very centre of the social and productive relations, transforming the instruments into the actual subjects and agents of production. Therefore, workers – as generic human beings – became just the 'living appendixes' of the machines – as Marx stated it – all scientifically and functionally distributed around them.

Because of mechanization, the skills or abilities required of people for taking part in production almost coincided with human beings' generic ones, and the market of labour and raw material expanded its limits reaching the peripheries of the world. Definitively, larger than ever social groups came to be involved, mobilized and mutually articulated: low skilled workers, women and children, migrants from all around the world participated in the overall productive process. Moreover, this second stage consisted of a different equilibrium between social autonomous movement, on one side, and order or fixing instances, on the other, characterized by a stronger and more concentrated capacity, for capitalists, to determine the sense and the articulation of the overall phenomenon. Otherwise, looking at the workforces, such a second stage of 'mobility' implied people's need to invent new forms of social and political opposition based on their physical concentration in the big factories, on the deriving possibility of communicating to each other their conditions and claims, and of attributing to their political action a sense of universality. It culminated with the Fordist and Taylorist epoch, and with its typical set of social relations, the analyses and explanations of which required the theoretical development of Structuralism. Foucault's concept of *discipline* – stemming from his prior inspiration to Structuralism – is very helpful in order to catch the inner sense, forms and procedures of this historical articulation of the productive factors.

Land and sea

It is particularly relevant, at this point, to refer to the work of another important author who gave a great contribution to the understanding and the description of the historical affirmation of capitalism. We are alluding to Carl Schmitt, who represented such a process as basically consisting of the dissolution of the political importance socially attributed to *land*, which for us means the crisis of the traditional conception and reality of rurality.

In the book titled *The Nomos of the Earth* and, especially, in the one named *Land and Sea*, Schmitt describes human history as characterized by a changing relation among men and the natural elements. In particular, *land* is the original source of each human sense of measure and ordainment, while *sea* had been ancestrally seen as an unsuitable ambient, source of dangers and reign of freedom, impossible to dominate. The relation between land and sea – and consequently between instances of order, stability and measure, and instances of freedom, uncertainty and hazard – changed when men, 'looking for the biggest whale', progressively developed the technical tools and proper abilities for facing and governing the oceans. In this way, they also matured the capacity to explore the world and to look at the land from the sea's wavy and changing perspective, which meant being progressively able to transform the social earthly ordainments by exercising the liberty of moving through the fluid element. Quoting Hegel (1820), Schmitt also argues that as for family life the condition is the stable ground of *land*, so the basis of industrial life – and therefore of capitalism – is the *sea*, i.e. the natural element which stimulates intercourse with foreign lands, being the support of movements and traffics rather than an element where sedentariness can be developed.

At a certain point of his fascinating historical and philosophical narrative, Schmitt considers also the intervention of a third element, *fire*, which in human history is the metaphor for the acquired dominion of technology. That meant the affirmation of big Leviathans on the world scenario, which the author identifies with the modern administrative States, as well as the complex, articulated, and mechanized procedures of industrial production. Their development implied a new configuration of the dialectic between freedom and ordainment, between social movement and functional fixation, and therefore the Leviathans' capacity to impose their order everywhere, to govern lands, oceans, and even air. This stage evidently corresponds to the Fordist and Keynesian epoch at the level of the social relations, and to the phenomenon of Imperialism at the level of international relations.

Due to his different political and philosophical approach, and partially as well to his chronological collocation – closer to us when compared with Marx – Schmitt's thought offers to our analysis a more explicit consideration of the violent abstraction modernity had made of the basic element of rurality, as well as of its ultimate *elementary* shift away from *land*. He also gives us account of the historical relevance and political consequences of the social struggle mediated by technology and enacted on the terrain of the dialectic between mobility and proximity, which continuously changes human relations with space.

Micro-physic and bio-politics

About such articulation of instances of order and liberty, and in order to better define and fully understand the mobility/proximity dynamic, we would like to take into account the work of Michel Foucault.

In our opinion, his thoughts represent an extraordinarily fruitful articulation of the terms of such a dialectic, and therefore provides the potentially most adequate framework to examine the evolution of essential phenomena which characterize the modern and contemporary world. His specific contribution, in relation to the two mentioned philosophies of history elaborated by Marx and Schmitt, consists of a new epistemological basis for the concrete terms we have used to define mobility and interconnectedness, necessary to functionally enforce the social instances of freedom and order – what we have so far figured as the spatially dynamical existence of people and the Leviathans. Moreover, for our specific aims, herein, Foucault's work and concepts represent a great methodological archive to be taken into account in order to elaborate new heuristic tools for the active interpretation of the social reality.

In the work published with the title *Power/Knowledge*, Foucault suggests that:

> we should direct our researches on the nature of power not towards the juridical
> edifice of sovereignty, the State apparatuses and the ideologies which accompany
> them, but towards domination and the material operators of power, towards forms
> of subjection and the inflections and utilizations of their localized systems, and
> towards strategic apparatuses. We must eschew the model of the Leviathan in
> the study of power [...] and instead base our analysis of power on the study of
> the techniques and tactics of domination. (Foucault 1980: 102)

Ultimately, Foucault's analysis emphasizes the ontological primacy of the 'micro-physical' architectures, techniques, and procedures of power, over its macro-level phenomenology or representation which we are used to identify with the State or the Capital. Nowadays, such a micro-physical dimension of power relations consists, therefore, of the mechanisms and rationalities which (Neo)Liberalism (its governing practices, rather than its ideology) put at the very heart of governance and of social and individual self-governance. Indeed, as Agamben (2007) and Lazzarato (2006) state, Foucault's analysis of the concept of power was based on the assumption that freedom is the social and individual general condition: actually, according to the French philosopher, 'there is no power without potential refusal or revolt' (Foucault 1988).

So, whereas Schmitt conceives *fixed ordainment* on one hand, and *freedom on the move*, or *movement as freedom*, on the other, as polarized opposite social and political conditions, fighting each other, according to Foucault they are just the two constitutive sides of the same human reality or sociological phenomenon called 'power'. It is thought of as a series of strategies acting within a net of relations, wholly coextensive to the social body, and always generically *productive*. Actually, for the exercise of power – so conceived – *human freedom* is the general presupposition and the contextual condition, while active and bodily *resistance* is its dynamical limit or obstacle, responsible for its procedural and ever-changing nature.

In order to better refine our conception of the mobility/proximity dialectic, thus it would be useful – following the Foucauldian approach – to distinguish the two complementary sides of modern exercise of power, namely consisting of *discipline* and *governmentality*. With the first term, Foucault indicates those 'dispositives' and techniques which positively transform and shape individuals' attitudes, postures and behaviours, by acting on the details of the human body and soul, in order to produce a specific ability, function, specialization, identity or subjectivity. Such techniques are properly named 'dispositives' by the French philosopher, since they literally *dispose* and *positively impose* specific conducts and forms to the human body and attitude. Eventually, discipline is that assembly of power techniques oriented to *fix* some specific feature and *order* to individuals' behaviour, with the purpose of conferring a particular sense or functionality or orientation to their *movements*. We could partially conclude that the product of the exercise of discipline – which acts out at an individual and punctual level – is the elementary facet of the social and historical articulation of mobility/proximity.

The other essential concept of the Foucauldian conception of power – beside discipline, and complementary to it – is *governmentality*. It is the specific mode of exercise of a historical paradigm of power, which he calls Bio-Power, now totally focused on and immersed into the sphere of human life: in fact it is human life, and its continuous enhancement, that is the object, the means and the goal of governing. Bio-Power corresponds to the historical emergence of an assembly of power's techniques and strategies consisting of the administration, management, regulation and orchestration of populations, life and their well-being. Such activities have in common an essential mechanism which consists of the effort of handling a complex interconnection of men and things, knowledge and nature, movements, obstacles and frictions. According to Foucault's definition of the concept, governmentality is based on science developed since the eighteenth century, and called *Political Economy*. Rousseau was the first who properly defined it as the set of knowledge and techniques concerned with how to apply and extend the mode of well-governing of the family house (etymologically, 'economy' stems from the Greek words *oiko* and *nomos*) to the public sphere of Politics (from the Greek *polis*) (cf. Larrère 1992). Therefore, such a new science must find out how to enhance the health and the wealth of the population, as the good father would take care of his family's well-being, prosperity and possessions. Actually, his role cannot consist (just) of a systematic, direct and detailed imposition of conduct – as the discipliner does – but he must rather use compassion, authority, and comprehension, in order to create, as much as possible, a suitable atmosphere of interaction, in which people can feel comfortable and act autonomously. Political Economy and Bio-Politics are strictly and mutually connected, on the logical and historical level. The emergence of a paradigm of power engaged in populations' life and health, as well as their continuous enhancement, are just the result of the same epistemological turn which fostered the development of the science and the practice of a public *paternal care* towards the modern *national* communities. Furthermore, it is to be noticed that in the historical development of the European Nation-States the biological fact of being born – in Latin, *natio* – acquires, historically and epistemologically, an ontological political sense. In fact, the modern concept and reality of *national citizenship* is emblematic of the historical fusion of the biological and political dimensions of life – as in the case of the familiar paradigm, typical of the *oiko-nomos*.

Actually, if Political Economy is the new science of reference for this novel paradigm of power, Foucault argues that *Liberalism* represents its practical declination or, more precisely, its *operational philosophy*. He defines it neither as a theory nor an ideology, but rather as an auto-reflecting practice, oriented to the accomplishment of some broad goals by the rationalization of the governing exercise. On one side, the latter must consist of a reduction of the government's positive enactment, and on the other, it must be rooted in the practices of self-governing implemented by the individuals, variously conditioned by morality and discipline, utilitarianism and respect of the general and universal rules stated by the legislator, and enforced by the police. In Foucault's conception, power and the process of governing become multi-centred phenomena, animated by many actors, motivated by heterogeneous motions, and rooted in many localities, each of which is characterized by different peculiarities, procedures and equilibria. Overall, from the governmentality perspective, the social body is not simply a static and passive object to be 'governed' from an external position – as in the classic paradigm of power – but rather an alive relational entity, whose intimate movements and interchanges could be funneled towards some partial or general goal, stimulated or discouraged, but neither stopped, nor punctually determined. So, from such a point of view, governing is a complex and plural phenomenon which actively involves individuals, institutions, motions, desires and things. Therefore, governmentality is rather the art of regulating the intensive self-generated flows of men, resources, energies, commodities and even values, peculiar of the modern and capitalistic age, by enforcing a net of circulation, by fixing some general rules, by trying to condition individual freedom throughout a *balancing* of incentives and penalties.

Whereas disciplines act individually, working out the intimate details, and creating specific attitudes, abilities, forms or functions in the human beings, governmentality is the exercise of techniques of power acting at a higher and broader level, interposing its 'dispositives' within those flows, modifying or adjusting social activities and instances, in order to potentiate the overall social corpus throughout it. In other words, the discipline is the art of positively fixing and fashioning the micro-details of the individuals' existence *in order to make them move productively* within the social body; governmentality is the art of strategically balancing *laissez-faire* and interventions within the overall social movement and dynamical activities in order to confer them with a general sense or finality, which is the enhancement of life. Finally, governmentality can be explained as the complementary concept to discipline, and it helps us to conceive the macro and collective dimension of the contemporary articulation of mobility and proximity.

Let us suspend at this point our analysis, and come back to the consideration of the historical steps of such a dialectic, in order to notice how the first two stages of mobility and proximity implied the typically modern phenomenon of the dissolution of rurality and of the traditional peasantry. Actually, they had been de-structured by mobility – we are referring to the phenomena of pauperization, vagabondage, proletarization – and transfigured in something else throughout the affirmation of the industrial forms of proximity or interconnectedness, as the new effective social ordering criteria – namely, urbanity, the modern proletariat and the division of labour progressively articulated around the big machines.

Indigenous migrants and relational machines

Within such a context, indigenous and native communities should be considered as exacerbated forms of *reserve* or *residue* of traditional rurality inside modernity, their being either violently constrained to seclusion, or induced to self-construct that condition as a form of resistance. They were literally a historical 'exception': their experience of modernity was characterized by a state of spatial and temporal 'immobility' and of a relational and historical 'isolation' – concepts which are clearly opposite to those of mobility and proximity characterizing, as stated, modernity and capitalism (Buscema 2005b). In terms of space, time and relations, their experience of modernity passed enclosed within territorial and social reserves, and fragmented in cluster-communities with very little possibility of even practicing exchange and interaction among each other. Though, more often their contacts were rather limited to violent struggles animated by the competition for the poor local resources – like in the specific case of the ethnic groups of the Mexican region of Oaxaca.

We can partially conclude by stating that indigenous people seemingly bypassed modernity and were bypassed by its essential dynamics. It is a paradox, since the social and political category of being 'indigenous' and 'native' indicates the persistence and bodily presence of a social group within a territory in spite of the historical events, and the ancestral connection of a social group with land, beyond chronology and facts. Such 'radicality' – here the term is used as a synonym of *rootedness* and, thus, of *immobility* – for indigenous people meant being absent, or just passive, in relation to the dynamics of history.

Such a paradoxical condition of *bodily presence and historical absence*, is fully consistent with one of the most important statements made by the Subcomandante Marcos. Actually, in occasion of the insurrection of the indigenous people of Chiapas, he pronounced these words: '*nunca más un mundo sin nosotros*', which means 'never again a world without us' [indigenous people]. Moreover, the relevance of this statement consists also of the expression of a novel will and need of those people of actively re-entering history, of trying to shape its ongoing course.

According to our hypothesis – structured on the basis of the Marxian and Foucauldian categories and methodologies, above analysed – such a phenomenon is now made possible by the current social and technological paradigmatic turn, which is animating and producing the third historical stage of the mobility/ proximity dialectic, within a context defined by many scholars as post-Fordist and/or post-modern (Sayer 1989, Harvey 1989, Negri 2008). Mobility and proximity, or interconnectedness, are now totalizing experiences, both in the extensive and intensive sense, virtually affecting each and every social group over the global world, each and every individual moment and aspect of the individuals' life. It must be highlighted that, given such a conception of those terms and of their extension, mobility and proximity cannot or should not be measured at all just in geographical or physical terms, but rather in terms of the relations and interconnections enacted by peoples and individuals, and intercrossing them.

All this is due to a new technological and social turn, consisting of the emergence of a global and globally networked society based on the historical availability of what we call 'relational machines' (Buscema 2005a). These new apparatuses are highly spread out within the social body, micro-sized, networking functioning, continuously innovated by social appropriation and usage, and increasingly tending to mediate all social activity and to be coextensive to the relational nature of the human being. Concretely, we are referring to cellular phones, the Internet, and all the articulated webs of services new ICTs provide, as well as to the current new degree of development of the means of transportation. It is necessary to clarify that such a process is not to be seen as technologically determined, but rather matured within the context of social micro-physical struggles, in the Foucauldian sense.

In comparison to the social-technological paradigm founding the modern big factories – as analysed by Marx – 'relational machines' become again 'media' in the etymological sense of the term, being tools, appendixes and means of expression and enhancement of the individuals and their faculties, their personalities, motions and instances. Whereas, at the very beginning of the process of emergence, productive application, and evolution of the modern machineries, Marx analysed how they were determining the dominion of the fixed capital over the workers, who were almost reduced to the condition of living instruments of the productive process.

In terms of collectivities, 'relational machines' therefore enhance the capacity of *social, horizontal, peer-to-peer relations* – what Foucault would call the micro-physical level of power – to contrast more structured and institutionalized forms of exercising power – what the French philosopher calls the states of dominion. Moreover, the difference is highly evident if we compare the two paradigms in terms of the social allocation of the technological terminals, in terms of their sizing and functions. Nevertheless, such a phenomenon cannot be conceived as an ultimate liberation of the social forces or a substantial escape from the conditions of subjugation and dominion. The ever dynamical dialectic of the power relations is already shifting towards the elaboration and the implementation of new forms of subsumption – based on governance practices and the strategies of financialization of society – specifically aimed at catching such new instances of contemporary sociality, and at funneling them into the mechanisms of capitalistic development (Buscema 2009).

Now, we will shortly refer to the results of the empirical research and the fieldwork activities we have carried out in the indigenous region of the Mixteca, in Sinaloa, in Tijuana (Mexico), and in the San Diego county in the US, from 2002 through 2005. In particular, we have focused our analysis on the migratory flows originated from a little indigenous community called San Jorge Nuchita, supplying workforce to the transnational agro-business labour market located on both sides of the border dividing Mexico and the US. It is important to stress that the relatively recent development of such massive flows of indigenous people from that remote area, and other ones of the Latin American continent, is to be conceived as the most patent sign of the current degree of extension and intensity of the mobility/proximity dialectic, which is one of the symptoms indicating the generalization of the (global) productive process.

Migration is now, for that region and their populations, an *ontological phenomenon* which is transforming and extremely re-sizing and re-shaping the relational status and resources of those social bodies and individuals. Especially, we have identified and studied two different migratory flows, both sharing the same geographical origin as stated – namely, the village of Nuchita – and the same economic destination or type of employment: the North American industrialized and intensive agriculture sector.

The first flow connects Nuchita to the North Mexican State of Sinaloa. The second connects Nuchita and the San Diego county, beyond the international border – famous worldwide because of the tragic frequency and violence of the deadly attempts of irregular crossings. Trying to drastically summarize, the two flows are different in terms of:

- how mobility and proximity or interconnectedness are socially organized: i.e., who actually manages and articulates the process of overcoming the physical distance between the contexts of origin and arrival; and
- where technologies and their use necessary to enact mobility and proximity are socially allocated: this means, in which point of the relations among enterprises and workers and indigenous communities' members these technologies are concentrated.

In the first case, the enterprise and its agents organize and structure both the chronological scanning and the spatial ambits within which this type of migration and such typically modern productive relations occur. The enterprise actually covers the costs related to the recruitment of groups of families from those remote communities, of their round trip back and forth, according to the economic needs of the enterprise itself and to the criterion of maximizing its own profits.

In the second case, instead, migrants themselves become able – which means 'relationally' able – to organize their journey, to figure out when, how and where to migrate, for how long, etc., by handling relational resources (namely, information, new languages and vocabularies, social contacts and spaces, ability to handle novel communication media), by using such relational machines to keep in touch with the recruiters, with their co-workers, with the 'coyotes' for crossing the border, etc.. Technologies and their usage, in this case, are not socially concentrated like in the former one, but socially spread out, involving the individuals, making them able and capable to get, share and process more information, to make their own decisions and to express their own motions, wills and needs – which are generally social, relational motions, wills and needs.

The functioning of technology, in the latter case, is not just passively suffered by people, but rather actively implemented before, during and after each migratory journey. Whereas, in the first case, technologies are used in order to objectify and subjugate people and workforce, in the second one they function as means of 'subjectivation' – namely, the process of becoming subjects and active actors.

The harvest of Dionysus

At the end of this chapter, and as a form of meaningful conclusion, it is necessary to clarify the sense of the title we have given to it. *Labor of Dionysus* (1994) is the enigmatic title of the first book written by Michael Hardt and Toni Negri together – the successful authors of *Empire* (2000) and *Multitude* (2004). In that book they focus their attention on the new threat characterizing the productive social forces within the post-Fordist epoch, consisting of the capacity of articulating relational spaces and combinations of powerfully productive forces, on the immanent plane of sociality. According to them, now this sociality tends to become politically and productively self-reliant: i.e. able to create better and richer articulations among 'social individuals' independently, or even in spite of, the ordering activity of the State and the instances of organization of capitalists. The way capital currently tends to reproduce and augment itself – the two authors specify – is more and more reliant on a secondary intervention in the productive activities, depredating part of the wealth autonomously created by the social body.

As stated earlier, according to the Greek and Roman mythologies, Dionysus is the divinity of the natural and mysterious energies making fruits grow and human pleasures and prosperity available – even though unequally distributed throughout society. Moreover, like natural energies tend to cyclically disappear because of the seasons, Dionysus is represented by mythology as persecuted during his peregrination around the world. Actually, he was also associated with *foreignness* and *madness*, since he was considered as a divinity inspiring and fostering the liberation of people from their 'normal' selves (we would say, with Foucault, *normalized* or *disciplined* ones). Thus, we argue that Dionysus can be interpreted as the emblematic figure of the *workforce* in the capitalistic system – both as a concept and a concrete condition of existence. As we have argued above, in fact, workforce is defined by its historically increasing contribution to the articulation of mobility and proximity – throughout vagabondage, migration and the other related phenomena which have constituted the raw and relational material of the capitalistic mode of creating value and, in more general terms, of enhancing social forces' productivity. Furthermore, the reference to (the *Labor of*) Dionysus permits us also to stress workforces' growing capacity to materially and technologically combine natural and mysterious energies in order to produce everything we (might) enjoy.

Thus, in conclusion, the new relational dimension of spatiality shaped by the current processes of articulation of mobility and proximity on a transnational plane – animated by the social forces and by their struggle – represents the inner premise to the phenomenon of the *creation of the world* as a global *urbe sine orbi* (Nancy 2003), or an unbounded social space. The formation of the *multitude* – as the social body and the set of productive forces adequate to globalization, and fostered by the autonomous initiative of subjectivities (Hardt and Negri 1994) – is the main result and, at the same time, the propulsive dynamic of such fundamental process which sociology must deeply explore. We should just finally consider that the observation of such phenomena from the extreme and *radical* point of view of the native-indigenous people and migrants of Latin America, not only corroborates the thesis above stated – according to which in the contemporary epoch they fully participate of the essential dynamic of generalization of the productive process – but it also demands the constitution of new forms of conceiving the access to rights, as well as to the wealth they contribute to produce. It demands a post-national citizenship and a post-waged labour remuneration, both to be rooted in the concrete innovative forms of life of contemporary global sociality, characterized – as said – by extremely profitable and increasingly autonomous expressions of mobility and interconnectedness.

Herein, by considering Marx's thought, we have stressed the new importance which these two processes or conditions gained for the social, economic and political relations since the settlement of capitalism. Through Schmitt, we have elevated those to the suggestive status of elementary forces which fostered the creation of the globe's entirety and dialectically constructed the historical *nomos* of the earth. Then, using the philosophy of Foucault, we have analysed mobility and interconnectedness as constituting the dynamic inner fabric of contemporary subjectivities, and highlighted how they implicate political conditionality and latency. Finally, the harvest of Dionysus is the metaphor of that *actual potential* of emancipation and strength engendered by the quite generalized faculty of communicating, producing and sharing wealth, and co-constructing alternative senses of development, based on the relational conquest of the energies of the *earth* – thus conceived not just as a nostalgic return to the *land*, but as the synthesis of the hybrid abilities prompted by the four elements humans can handle.

References

Agamben, G. 2007. *Che Cos'è un Dispositivo.* Roma: Nottetempo.

Buscema, C. 2005a. *Camminare Producendo. Le Migrazioni dei Braccianti Mixtecos dell'Industria Agricola Nordamericana.* Soveria Mannelli: Rubbettino.

Buscema, C. 2005b. I migranti mixtecos tra relazioni produttive moderne e postmoderne, in *Le Migrazioni tra Ordine Imperiale e Soggettività*, edited by G. Sivini. Soveria Mannelli: Rubbettino.

Buscema, C. 2007. *bOrder-Disorder. Presupposti d'Analisi della Riforma Migratoria degli Stati Uniti d'America.* Cosenza: Quaderni di GAO, Collana Ricerche.

Buscema, C. 2009. *Tempi e Spazi della Rivolta. Epistemologia Critica delle Soggettività (Migranti) e dell'Antagonismo ai Tempi della Governance e della Finanziarizzazione.* Roma: Aracne.

Foucault, M. 1980. *Power/Knowledge: Selected Interviews and Other Writings, 1972–1977.* Brighton, Sussex: Harvester Press.

Foucault, M. and Perrot, M. 1983. *Panopticon, ovvero la Casa d'Ispezione.* Venezia: Marsilio.

Foucault, M. 1988. *Michel Foucault Politics, Philosophy, Culture: Interviews and Other Writings 1977–1984.* New York: Routledge, Chapman & Hall.

Foucault, M. 1994a. *Poteri e Strategie. L'Assoggettamento dei Corpi e l'Elemento Sfuggente.* Milano: Mimesis.

Foucault, M. 1994b. *Dits et écrits.* Paris: Èditions Gallimard.

Hardt, M. and Negri, A. 1994. *Labor of Dionysus: A Critique of the State-Form.* Minneapolis: University of Minnesota Press.

Hardt, M. and Negri, A. 2001. *Empire.* Harvard: Harvard University Press.

Hardt, M. and Negri, A. 2005. *Multitude: War and Democracy in the Age of Empire.* Harmondsworth: Penguin.

Larrère, C. 1992. *L'invention de l'économie politique au XVIIIè siècle. Du Droit naturel à la physiocratie.* Paris: PUF.

Lazzarato, M. 2006. Foucault oltre Foucault. *Multitudes.* [online] Available at: http://multitudes.samizdat.net/Foucault-oltre-Foucault [accessed: 28 March 2008].

Marx, K. 1887. *Capital: Volume One.* [online] Available at: http://www.marxists.org/archive/marx/works/download/Marx_Capital_Vol_1.pdf [accessed: 18 February 2010].

Marx, K. and Engels, F. 1848. *Manifesto of the Communist Party.* [online] Available at: http://www.marxists.org/archive/marx/works/download/manifest.pdf [accessed: 18 February 2010].

Moulier-Boutang, Y. 1998. *De l'Esclavage au Salariat. Èconomie Historique du Salariat Bridé.* Paris: Puf.

Nancy, J.L. 2003. *La Creazione del Mondo o la Mondializzazione.* Torino: Einaudi.

Polanyi, K. 1944. *The Great Transformation. The Political and Economic Origin of Our Time.* New York: Farrar & Rinehart.

Schmitt, C. 1996. *Land and Sea.* Corvallis OR: Plutarch Press.

Schmitt, C. 2006. *The Nomos of the Earth in the International Law of Jus Publicum Europaeum.* New York: Telos Press Publishing.

PART II
Discourse/Identity on Proximity and Mobility

Chapter 4
The Semiotics of (Im)mobilities: Two Discursive Case Studies of the System of Automobility

Chaim Noy

Introduction: textual auto(im)mobilities

This chapter explores two different yet related illustrations of textual (im)mobilities within the system of automobility. The term textual (im)mobilities relates to the occurrence of texts within systems of mobilities and to the ways that these texts become meaningful by using the unique resources that these mobile systems afford. In other words, this chapter investigates what happens when texts and (im)mobilities meet. The chapter argues that (im)mobile textual performances are accomplished, and that these performances concern the semiotics of texts as much as they concern the semiotics of (auto)mobilities.

The notion of (im)mobilities is explored through attending to mobile and immobile discourses that are part of the 'system of automobility' (Urry 2004), itself one of the most complex and under-researched mobility systems of late-modernity. As sociologists of mobility argue, automobility is the 'avatar of mobility' (Thrift 1996: 272). Automobility studies have rapidly developed during the last decade, suggesting insightful analyses into the formative role automobilities play in everyday life (Featherstone, Thrift and Urry 2005). The field employs various mobility sensibilities and sensitivities, dedicated to the empirical and theoretical study of mobilities, immobilities, and related concepts (proximities, connectivities, motilities, etc.), and of their consequences on contemporary societies. Specifically, works on automobility suggest a twofold appreciation of this concept whereby it is viewed simultaneously as a research paradigm (epistemological perspective) and as a social condition (a central characterization of sociality in late modernity).

The first case of textual occurrence to be explored concerns bumper stickers, which are short textual notices that are temporarily attached onto the external (rear) surfaces of cars. The second case concerns personal monuments which were erected in the memory of people who had died in car collisions, and are located by the side of the road. These cases illustrate textual mobilities and immobilities, as they ostensibly occupy two ends of the (auto)mobility continuum: while bumper stickers are generally viewed as cases of mobilized discourse, i.e. texts which follow the cars' mobilities, personal monuments are generally viewed as cases of immobility, i.e. discourse that is inscribed onto monuments of stone and therefore unmovable (physically) and immobile (symbolically). Combined, these auto(im)mobile case studies suggest that the dialectics of mobilities and immobilities within automobility, are the *basic and defining features* of these public textual occurrences, and of the ways that they become socially meaningful or performative. These case studies also tell a story about the semiotic processes that transpire within automobility and about the politicization of mobilities.

Furthermore, both cases of textual occurrences assume meaning only under the conditions of *visibility* that are available in the system of automobility and are characteristic thereof. The notions of textual (im)mobilities and visibilities are linked, because visibility is what grants the spaces of automobility their status as public spheres, because visibility allows mobilities to be observed and detected, and because – with particular reference to textual occurrences – visibility suggests a degree of proximity by which reading (deciphering) these texts, or at least recognizing them as discursive occurrences, is possible.[1]

The present work furthers earlier investigations on the performances that texts inhabit in mobile socio-material systems. Earlier research probed commemorative visitor books that are located in highly symbolic sites, conceptualizing these books as an immobile platform located in a matrix of global travel, tourism, and mobilities (Noy 2008a). In another work, spoken (rather than inscribed) discourse was examined as it transpired in, and as part of the activities of, driving a car and of riding inside one (Noy 2009, forthcoming). These works share an ethnographic appreciation of (im)mobile (con)texts, and of how within these (con)texts and through the resources of (im)mobility meaning(s) is (are) created and communicated. Hence this chapter will try to 'read' texts as they extend unto the material (and other) realms of mobilities and immobilities, and suggest that they would remain quite meaningless if stripped away or decontextualized from their ecological and mobile 'place of being' (Heideggerian *Dasein*) or 'place of being on the move'.

As in earlier works in this field, this chapter's conceptual point of entry into the discussion of (im)mobilities of texts and the meanings they perform, concerns first and foremost the appreciation of the material dimensions of the texts, or the *materiality of texts*. This term denotes the material qualities of texts of different sorts, suggesting that these texts are not abstract semiotic entities. Texts make meaning insofar as they offer a decipherable code (i.e. language), which is implanted, engraved, written, inscribed, smeared, etc., in and on objects and surfaces, and is thus part of the actual social world. The appreciation of the materiality of texts and their embodied state in society, which is a post-structuralist appreciation, is presently inspired by two lines of research and theory. The first concerns works which evince a material, situated and interactional appreciation of discourse (see review in Noy 2008b); the second builds on Latour's (1987)

---

1   It is no coincidence that one of the leading sociologists in the field has also been influentially writing about visuality (in the context of 'the tourist gaze', see Urry 1990).

theorizing, and specifically on his view of the role documents play in science and in society.

(Auto)mobile methods

The research on which this chapter reports is part of a larger examination of semiotics and discourse in automobility. The methods used in this project are varied and are suited to fit the essentially mobile nature of the field (on 'mobile methods' see Noy forthcoming; Urry 2007: 39–42). Rather than 'freezing' social reality in order to 'experiment' on it, as traditional sociology has done, automobility research shows empirical sensitivities complemented by epistemological sensibilities to mobilities and related concepts.

The present research builds on ethnographic observations of and interviews with a range of social actors in automobility, that were conducted during the years 2004–9. These social actors include professional and laymen car drivers (57), passengers (in both private and public transportation), pedestrians (27) and car mechanics (6). With specific regards to the issues of bumper stickers (BSs) and roadside death monuments (RDMs) additional research was pursued in the following extent: 27 brief, on-road interviews were conducted in order to learn about BSs and related automobile practices. These exchanges took place in parking lots and in urban traffic lights, where moments of immobility were seized in order to afford interaction with drivers. The interviews (and accompanying photos) are usually taken from *within the author's car* during everyday traffic interactions in the city of Jerusalem (Israel). Since the research takes place *in situ* – that is on the road, special attention was paid to issues of safety, so that these interactions do not put anyone (interviewees and researcher in their capacity as pedestrians, passengers and drivers) at risk.

With regards to the study of RDMs, discursive and semiotic analysis of 58 RDMs (located mainly in rural and southern parts of the country) was conducted, complemented by 11 telephone interviews with social actors who took part in their construction (usually relatives of the deceased). Finally, as automobility pervades our everyday lives, my personal experiences and contemplations in and of automobility in Israel have also contributed to this chapter.

Bumper stickers: the mobilities of textual objects

In Israeli culture and especially in its political culture BSs are a celebrated phenomenon. The widespread use of BSs has received popular attention in the form of public discussions in the media and in popular songs, where the intensity of their use and their oftentimes nationalist contents have been discussed.[2] BSs have also been researched, primarily in the fields of communication, discourse studies and folklore (Bloch 2000a, 2000b, Livnat and Shlesinger 2002, Salamon 2001). Reported findings indicate that between 30% and 80% of cars in Israel carry political BSs (Bloch 2000b: 435, Salamon 2001: 117). The variation may result from different periods when the samples were taken (in heightened political times there is a noticeable rise in the use of BSs), or from different urban locations of sampling, reflecting degrees of political involvement and forms of expression. However, even the lower end of the figures indicates rather high percentages of BSs usage on 'regular', everyday basis. Note that the figures refer to political BSs defined narrowly, that is to BSs that include explicitly political expressions. If BSs that are ideological in the larger sense are taken into account, the figures are even more impressive. Indeed, in what follows a broader definition of the political sphere is accepted, where various claims of identity and cultural preferences are also viewed as political expressions.

Bloch (2000b: 434) correctly traces the origin of the use of political BSs in Israel to the late 1970s, with the 'Peace Now' BS, which represented the Leftist political movement of the same name (Shalom Achshav in Hebrew). The widespread use of BSs followed, reaching the high figures that are reported in the research during the mid-1990s, following the major event that was the assassination of Prime Minister Yitzhak Rabin (1995). Further research has revealed how political meanings are communicated in Israeli's public sphere, how these short messages are discursively constructed, and how they assume meaning through intertextual references (such as paraphrases of contents and forms of earlier BSs). Yet it is worthwhile noting that in these multi-disciplinary contributions, very little reference is made to actual (im)mobilities, to the underlying infrastructure that enables them, and to semiotic implications that they carry within the realm of automobility.

As indicated earlier, the lead that facilitates the (re)conceptualization of the charged scene of BSs in Israel in terms of (im)mobilities concerns the *materiality of the actors* in the scene. Material sensibilities and sensitivities supply new perspectives into how meanings are created, sustained and negotiated in the local auto-BS scene. Hence in what follows, BSs will be conceptualized as 'textual objects' and not merely as 'texts'. This conceptualization will outline BSs' (im)mobile relations to other objects and their emergent meanings.[3]

The first two material aspects to be noted do not concern BSs themselves, but the materiality of cars, which are the *physical carriers of BSs*. Consider, for

---

2   I am referring in particular to the 'Sticker Song', which was written by one of Israel's foremost novelist, David Grossman, and performed by the famous rap group, the *Dag Nahash* (2004).

3   The literature offers a number of concepts describing the juncture of discourse, materiality and mobility, such as 'textual artifacts' (Silverstein 1996), 'graphic artifacts' (Hull 2003) and 'hybrid inscriptions' (Noy 2008b). I presently use the term 'textual objects'.

instance, Bloch's (2000b) account of the origin of political BSs in Israel, which lies in their widespread use by the Peace Now movement in the 1970s. Bloch accurately points out that BSs amount to a public medium of communication which is inexpensive, and which offers access to the public scene and is readily accessible to many citizens (Bloch 2000b: 443). However, in line with material considerations it should be noted that while BSs are admittedly inexpensive to produce and are usually distributed freely, *cars* are not so cheap and are usually not distributed freely. Hence, arguing for the wide accessibility of BSs evinces a naturalized (unproblematized) appreciation of the sphere of automobility and of the fact that what makes meaning is not 'texts', nor even the objects of BSs, but the *physical juxtaposition* of the textual object of the BS with the object of the car. Discussing BSs in relation to the leftist Peace Now movement (or to other social and political organizations), which initially used BSs in the late 1970s, remains incomplete without adding a neo-Marxist material sensitivity, which suggests the term 'bourgeoisie' be taken into account. Indeed, the leftist Peace Now movement consisted mainly of Ashkenazi Sabra Israelis (Israelis who were born in Israel but were of European background), of affluent quarters of (Jewish) society. This addition perspective sits nicely with the fact that mobilities of various types, and their accessories – cars and cars' BSs in this case – have always been a matter of privilege and of those privileged enough to enjoy them (cf. Urry 2004: 26).

Another contribution that material sensitivities give to the exploration of automobile BSs concerns the *volume of traffic*. While the figures mentioned earlier regarding the prolific spread of BSs in Israeli society are impressive, in the material context of automobility there is a missing link that concerns the density of the traffic and consequently the density of the *occurrences of BSs*. Consider that the local traffic congestion in Israel is approximately 128 vehicles per kilometre of road (or VKR) (measured in 2005–6).[4] This figure varies considerably across countries. For instance, Belgium, France, New Zealand, Poland, Spain, Syria, and the United States range between 20–40 VKR; Jordan, Mexico and the United Kingdom range between 60–90 VKR; and Germany and Hong Kong, China have above 210 VKR. These figures affirm the intensity of BSs on Israeli roads. Yet the point is not the specific figures, but the fact that the estimated percentage of BSs on cars means something quite different for the *actual experience* of people interacting within different systems of automobility. Without accounting for the material environment – car congestion, in this case – only a partial picture can be depicted with regards to the semiotics of (auto)mobility. The point is, again, that if BSs are detached from their physical carriers and their physical environment, the picture of how effective and meaningful they are is necessarily partial.

While these aspects attended to the materiality of cars, we now turn to the materiality of the BS itself, insofar as it is conceptualized as a 'textual object'. If BSs are objects of sorts, then according to Appadurai's (1986) influential thesis, they have a 'social life' that consists of spatial trajectories and paths of circulation

---

4   See Hamadi and Chittajallu 2008: 132–5.

that intersect with cultural, political and ideological semiotic grids. BSs are not merely 'texts' that are parts of 'systems of representation'; rather, they carry a value that is related to their quality of 'object-ness' and to their paths of circulation and mobility. This discussion enlarges the scope of mobilities with regards to BSs by addressing the mobility they embody over and above the mobility that is granted to them once they have been attached to cars. In other words, BSs' (im)mobility on cars' surfaces is only *one part of the overall mobility they enjoy*, and as we shall see, this fact is consequential in terms of the meanings they perform.

Consider how BSs arrive into the system of automobility. Paths of distribution of objects suggest various socio-spatial mobilities. Interviews with car owners and drivers indicated four major categories of distribution by which they attained their BSs. There are two modes of mass distribution of BSs and two modes of restricted or exclusive distribution. The first of the two modes of mass distribution concerns BSs' mobility along routes of distribution of mass newspapers. This form took place on a number of major national/political events (such as the Israeli invasions into Lebanon and Gaza), when BSs were coupled with the most widely distributed newspapers in Israel. At those times, anyone who bought these newspapers would have freely received stickers. This form of mass distribution is impersonal and accounts for some of the findings reported above, regarding the high percentage of BSs on Israeli cars.

Sixteen (of 27) interviewees indicated a second mode of mass distribution, which is of a personal nature. This mode concerns the distribution of BSs at urban junctions as part of national political campaigns. Recently, this mode of distribution was performed effectively by youths, who were mobilized by the nationalist Right wing movement which objected to Israel's withdrawal from the Occupied Gaza Strip (during August, 2005). As part of this campaign, the distribution of BSs at urban junctions proceeded on an everyday basis for a number of weeks, making effective use of the fact that cars must come to a stop when waiting for the green light, thus enabling pedestrians to approach them. In the terms used by Urry (2002), cars' temporal immobility offered occasions of 'co-presence' of pedestrians and automobiles, whereby the former could interact with the latter. As the campaign was intensive, nearly all drivers in the city of Jerusalem encountered these youths at one point or another. This implied that what is of interest is not so much the cars that had BSs, but rather those which *did not*. In light of the pervasive campaign, it became clear that cars which do not carry BSs are *saying something*; and that they belong to drivers who actively and repeatedly refused to put an anti-withdrawal BSs (which were usually also racist), on their cars. Hence, under conditions of mass distribution, sometimes the *lack of a BS* is indicative of an ideological stance and as such marks the vehicle that is BS-free.

While these modes of mass distribution reflect major events and political campaigns, most of the interviewees (21) indicated that they attained their BSs via limited modes of distribution. Two different types of sources were mentioned, both of which are of an exclusive nature. The first source concerns particular events which car owners (or someone related to them) had attended. The events that were mentioned included a number of military ceremonies (the completion of infantry basic training phase, parachute course, and the like), political rallies and demonstrations and sport events in which BSs were given out freely. The second source representing an exclusive mode of distribution concerned particular places and attractions where BSs were offered (freely or with a charge). A visit to a Che Guevara Museum in Bolivia was mentioned, and so were visits to wineries in the Northern parts of Israel, to nature reserves, to opening parties for a number of local restaurants and bars in Jerusalem and welfare fundraising events.

Figures 4.1 and 4.2a supply illustrations of phases in the 'social life' of the textual object of the BS, as these take place on and off the road. Figure 4.1 presents the living room table in my apartment, during a political meeting with a candidate of one of the parties running for the parliament (the meeting took place on January 21st, 2009, three weeks before the general elections in Israel). The activists of the *Hadash* party brought various promotional items (pamphlets, shirts, notebooks, etc.), which were freely offered to those attending the event. Among them, the *Hadash's* red BS was also offered (appearing in the centre, near the empty coffee cup). The sticker has the party's name on it in both Hebrew and in Arabic, the text 'Building a New Left' (in Hebrew), and the party's webpage and icon. Note that the short text is a word play, because the Hebrew word for 'new' – *hadash*

Figure 4.1   Prepared for mobility: political BSs on living room table

Figure 4.2a   (Auto)mobile traces of Shmeltzer's concert

Figure 4.2b   'I'm a Simple Jew'

– is pronounced exactly as the party's name. Also note the various promotional materials are mobile in different ways, such as shirts that are meant to be worn, and evoke different mobilities that require visibility.

Figure 4.2a is a photo that was taken near the Jaffa Gate by the Old City of Jerusalem in July, 2007. On the rear of a white Citroën, under the window, a white BS is pasted. The BS included two sentences, the first of which is larger and in Hebrew: 'I'm a Simple Jew.' Below, in smaller letters and in Yiddish (written in Hebrew letters): '*H'bin a Poshiter Yid*' (meaning the same as the Hebrew text). In the brief automobile interview, the driver, an Ultra-Orthodox Jew in his 40s, indicated that the sticker refers to an Ultra-Orthodox Yiddish pop star by the name of Lipa Shmeltzer, who recently released an album by this name (*H'bin a Poshiter Yid*). The driver got the BS at Shmeltzer's live concert.

Note that bumper stickers are only rarely located on cars' bumpers. Their preferable location is higher, in a more visible spot, usually on the rear window. This detail is in line with the appreciation of automobility as a highly communicative sphere. As Featherstone (2004: 8) notes, '[t]he automobile becomes a new form of communications platform with a complex set of possibilities', which include, on a basic level, 'the windscreen, windows and mirrors to the inter-automobile moving figuration of cars, and involve interactions and modes of presentation of the auto-self to others in the temporary "fluid choreography" of the shifting reference group of traffic'. It is within these visual affordances that BSs may flourish as they do. As indicated in the introduction, texts embody the notion of proximity because they are understood as codes that require up close visual attention in order to be deciphered. One of the local BSs humorously plays on this notion, as it read: 'If you can read this, you are too close!'

While the four modes of distribution described above offer socio-spatial paths through which BS move into and enter automobility, there are various modes through which BSs exit the system. Interviews with car owners, car mechanics and my own observations reveal three different types of events that are responsible for the removal of BS from cars, and thus for the completion of their 'social life'. The first and most pervasive event is that of the car wash (whether commercial or personal). Drivers indicated that when their cars are being washed they specifically remove BSs, either because these stickers are irrelevant (the event for which they were produced has passed, for instance), or because they were worn out and did not look nice or perform their communicative task effectively.

Two additional modes of removal of BSs were mentioned. The first is the active removal of BSs from cars by pedestrians who apparently did not agree with the sticker's message. From the experience of car owners with Leftish political BSs in Jerusalem (myself included), these BSs' are in use on car for little more than a few weeks before they are pealed off (sometimes with damage purposefully done to the car). Here, too, pedestrians interact with cars, making use of the fact that cars are at rest and that they are empty and not protected.

The second mode was indicated by Palestinian car mechanics (tinsmiths employed in East Jerusalem). This mode concerns cars that are sold by Jewish settlers in the Occupied Territories to Palestinians. Since settlers' cars typically carry a number of political BSs, Palestinian auto tinsmiths indicated occasions when, after purchasing a car from settlers, their new Palestinian owners arrive at the garage asking for the removal of the stickers, in addition to the removal of other signs that indicate the ideology of the car's previous owners (such as the double shielded and darker windows, that are installed in settlers' cars in order to prevent injury from stones thrown at them). In other words, a re-doing or re-dressing of the object of the car takes place in these tinsmiths' garages, transforming the semiotic object of the car, enabling its mobility in different socio-ideological and national systems.

The socio-spatial paths that BSs travel on their way into automobility (distribution) and out of it are consequential in terms of the social meanings they perform. BSs enjoy mobility and have a 'social life' prior to their appearance within the system of automobility. This fact enhances the field of exploration of BSs, and the meanings that can emerge from these mobilities. These mobilities are nested within and built on *social networks and social travel*. In other words, there is a whole semiotic dimension to BSs which concerns their social mobilities prior to – and of course also during – their automobile phase. In this vein, consider the *Hadash* BS awaiting automobility on the living room table (Figure 4.1). This point in the life of the BS is a nodal one, because it lies – temporarily unmoving – between distribution and usage. BSs of this type can be traced back to particular social events and particular people who attended them and share similar social circles, class and ideologies. The same is true for the 'I'm a Simple Jew' sticker (Figure 4.2b). First, the driver's account was illuminating because the meaning of the sticker could not be inferred from its text. In fact, before the interview exchange I had thought of the text as one which belongs to Jewish racist BSs, many of which are visible on Israeli roads. Instead, the text indexes a very different discourse, one that concerns issues inside Ultra-Orthodox society, where it performs a subtle criticism by celebrating the common person as opposed to powerful rabbinical figures.

Second, what the driver verbally explained is implicated in the performance of the sticker as a material and mobile object, which is to say that it was given by someone in particular to someone else in particular at a particular (exclusive) social event. Of course, I did not know of Shmeltzer's concert, but this is precisely the point. Through circulation in an exclusive manner and within particular social networks, a crucial part of the semiotics of BSs is denied from general audiences. While research on BSs in Israel (above) stresses the popular and shared meanings BSs enjoy, the findings of this research indicate that meanings are shared – can only be shared – within particular social milieus, and there are degrees of implicitness of meaning that are concealed in the public sphere.

BSs' mobile meanings are therefore by no means limited to the meanings that are expressed through their content, or even to the intertextualities that they evoke. BSs do not amount to simple (textual) expressions in and of themselves, but to *mobile and material traces of events, places and peoples* (Noy 2007, 2008a).

These events are meaningfully deciphered by their selected audiences. Note that since a quarter of the cars examined in this research carried more than one BS (the remains of a faded green BS under the right window in Figure 4.2a can be discerned), these surfaces can be said to be telling of a number of events of which traces are co-present.

At some point during their social lives – after entering automobility and before exiting it – BSs and cars are coupled. This juncture is accomplished by the cohesive paste on the back of the BS, which, in socio-material terms, is an agent that accomplishes the temporal association of two separate objects. This conjuncture is also a case immobility, because in order to assume automobility BS must be well fastened unto cars' rear windows. Once on the car, it would be inaccurate to view the car as an 'object which has a sticker attached to it', inasmuch as it would be inaccurate to view the sticker as an 'object which has a car connected to it'. Neither view is accurate because the pasting of the sticker unto the car creates a new automobile actor: the BS-car. The new social actor amounts to a gestalt; to a whole that is more meaningful than the sum of its parts. The BS-car is a *discursive, social and oftentimes also political* vehicle. It embodies temporalities, such as those that are referenced via the paths and nodal events of the circulation of BS, and social networks. While the methods that were employed in this research did not allow the following of the routes and mobilities of the BS-car hybrid, it is reasonable to assume that they too follow paths of social networks through which BSs are mobilized (travel to and parking in workplaces, leisurescapes, schools, and to friends, etc.).

Textual immobilities: roadside death monuments in Israel

The second case study that illustrates the conjuncture of semiotics, texts and auto(im)mobilities is roadside death monuments (RDMs).[5] Commemoration monuments emblematically represent an attempt to capture and arrest activities and mobilities that are associated with life and living, and thus 'freeze' the memory of departed people or past events. Specifically, RDMs are erected in the memory of people who were killed in traffic collisions. They are typically located between 5–15 metres off the road, in a visible location, near the actual site of the collision. Their physical construction routinely includes a large stone on which personal information and the circumstances of the fatal collision are depicted. Oftentimes, icons and symbols that index automobility, the collision and/or the deceased are also included.

RDMs bring to the fore one of the incredibly dear prices that societies pay for automobility. Yet the problematic issue seems to involve not merely the endemic

---

5   I adopt the term RDM (following Reid and Reid 2001), from a number of terms that are used in the literature, because I reject the use of the term 'accident' (implying incidental causality).

collisions themselves, but their denial and erasure by powerful actors within the 'regime of automobility' (Böhm et al. 2006). As Featherstone (2004: 3) notes, the collision is denied,

> because it is not seen as a normal social occurrence, but more as an aberration. The victims are dispatched to the hospital, the car to the repair garage or scrapyard and the road is quickly cleansed of traces of the crash by the accident services and the 'normalcy' of traffic flow restored.

In the literature on RDMs, Henzel's (1991) work in North-East Mexico is notable, arguing that RDMs are complex semiotic constructions that embody secular commemoration. Later research examined various aspects, ranging from the motivations for the construction of RDMs, through the ideologies and meanings that underlie them and the concrete fashions by which commemoration is practiced, to comparative analysis (see reviews in Clark and Franzmann 2006, Hartig and Dunn 2002). The literature accounts for the emergence of RDMs in the 1980s–90s by (postmodern) tendencies for individualism in Western and Westernizing societies. The RDM is erected in order to tell a story of an individual, but as a social phenomenon it also tells the story of individualism. RDMs embody subversive and resisting voices, which challenge the erasure of the memory of those who were killed. The grassroot (non-institutional) type of activity that is involved in the construction of RDMs evokes a protest against normative and canonic types of commemoration. Oftentimes, these voices resist normative religious rituals of commemoration, usually in the form of Christian doctrines. RDMs also challenge the capitalist order, where the smooth and ceaseless operation of automobility is a crucial factor, regardless of the toll that societies must pay. RDMs 'endeavour to inscribe the site as a place of tragedy and remembrance, by refusing to erase the incident from public memory and allow drivers to relax back into the normal traffic flow' (Featherstone 2004: 3).

In Israeli automobile culture, RDMs became salient in the early 1990s as well – as were political BSs – as a result of the globalization of ideologies of individualism and privatization, and the accepted ways of celebrating them. Although the number of RDMs is continuously growing (there are presently about 1,000 RDMs in Israel), the only research conducted on the topic to my knowledge is Vardi's (2006) thesis focusing on personal versus collective-national memorialization. This is probably due to the historic hegemony of national and military memorialization in Israeli culture.

The first step in attending to the semiotics of immobility is acknowledging the *discursive nature* of these monuments. Memorials can perform effectively without linguistic engravements: various graphic icons and symbols can fulfill the goal of indicating what/who is/are commemorated. The case with RDMs is different because they are built in order to commemorate the *personal memory* of the deceased. This act necessarily concerns her or his personification, of which the foremost way is the evocation of the person's name. In the age of liberalism it is the name and the autobiographical narrative that the name indexes, that supplies the common means of referencing individuality. RDMs' raison d'être is granting the deceased an individual identity, vis-à-vis the quantification of automobile fatalities in official records and in the media and the anonymous character of traffic (we usually do not know the names of people in cars). For this reason it is essential for the semiotic efficacy of RDMs that at least the deceased name will be specified. This is indeed the case with all the RDMs that I have recorded. While oftentimes RDMs tell a larger and more detailed story, the minimal discursive unit includes the deceased name(s), even if only the personal name.

RDMs effectively use the semiotic resources that are available in auto(im)mobilities. They do so via the use of one of the most emblematic embodiments of immobility, i.e. the memorial (or commemorative) monument, as it is located in one of the most emblematic embodiments of mobility, i.e. the system of automobility. Since traffic and the passing of cars are one of modernity's recognized symbols of fleetingness and evanescence, RDMs evoke the drama of immobility and the arresting of time in the heart of this foremost site of mobility.

In terms of signs in automobility, RDMs – like the traffic sign system – are reflexive in that they refer their viewers *back unto* automobility. RDMs do not invite people to purchase commodities nor to vote for politicians, but, like signs that indicate that a curve in the road is approaching or that a particular speed limit is set, they have an *event of automobility* as their reference. Uniquely, this event lies in the past and not in the future, thus running against the progressive (modern-liberal) ideology that automobility promotes. Put again, RDMs index an event that transpired on the road, and in this sense they tell a story not only of an individual person (or individualism), but also of an event of automobility – a constitutive event at that – the fatal collision.[6] RDMs can therefore be read as *documents* of a particular type, which engage in and offer an account of fatal automobile collisions. While it is clear that the conscious motivations behind the construction of RDMs revolve around the commemoration of those killed, their genre can also be understood as a semi-institutional document, located in the grey area 'between-and-betwixt' the formal system of traffic and personal commemoration. This document offers more than commemoration; it also offers ascriptions and the distribution of responsibilities with regards to actors in the scene of automobility.

When viewed in this light – as documents of this genre – the texts engraved on RDM interestingly correspond with another category of automobile texts, which also describe the events of the fatal collision and which also offer (in addition to descriptions) ascriptions regarding causes and responsibilities. Beckmann's (2004) intriguing work on the bureaucratic discursive reconstructions of automobile collisions, argues that the collision disrupts traffic and causes the dismembering of cars and drivers, which must be *reassembled by experts* in the form of reports that are produced in the aftermath of the collisions. With regards to fatal collisions,

---

6   See Brottman (2002), which is dedicated to the chilling cultures and aesthetics of car crashes.

Beckmann (2004: 92) rhetorically asks: '[t]he common journey of car and driver has ended with the accident – or has it just been interrupted?' In order to restore the physical flow of traffic and simultaneously in order to justify its price, the actors in the scene of automobility are transported (mobilized) away from the road (the physical scene), to various other theatres, such as courthouses, hospitals, police offices, laboratories and so on.

Beckmann (2004: 93, original emphasis) concludes that:

> '[w]ith the completion of the report form, the police officer automatically completes a *homogeneous* reality in which experts have agreed to take certain accident characteristics for "real" and "true" – *detached from personal interpretation.*' The objectivist and universalist genre of formal agencies' reports can be viewed as discursive performances which confront, but to some degree also complement the discourses that are performed via the RDMs. While the former establish the collision's 'objective truth' ('expert knowledge') the latter establish its 'subjective truth' ('lay knowledge'). Both, however, are engaged in power-struggle over sovereignty, knowledge and representation. Formal reports 'enable the governance of very specific social conflicts, labelled accidents. The classification and documentation of crashes is one central policy tool to reproduce a traffic system that has been, and still is, subject to a variety of controversial mobility views'. (Beckmann 2004: 95)

Surely, RDMs emerge as an alternative textual 'policy tool' and view, which (re)evaluate the fatal collision.

Figures 4.3 and 4.4 illustrate the play of (im)mobilities in RDMs' discourse of car collisions. These specific illustrations are located by the side of the highway on which I travel on a weekly basis on my way from Jerusalem to the college where I teach (located near the southern city of Sederot. These are two of six RDMs that I pass by on my way). These RDMs are located right off the road and are easily noticeable. While the arguments that tie their discourse to automobility are sometimes implicit and sometime explicit, the very location of the RDM in physical proximity to the road positions them within the semiotics of the automobility network. Their performance is thus accomplished indexically.

In addition to the spatial proximity to the road, the Ivannovs' RDM (Figure 4.3) accomplishes a meaningful act by textual and iconic reference to the collision and to automobility. In the top of the black plate an image of a (populated) car is describable. The image might be a reproduction of the Ivannovs' actual car, or an iconic reference to an automobile as such. In any case, the icon restores the actors – both car and people – to the state they were in prior to the crash (the 'ideal state'). Beneath it, the text, too, ties the monument to the event of the crash. The upper part is written in Hebrew: 'To the memory of Vitali and Natalya Ivannov who

Figure 4.3   'Forever with us': the Ivannovs' RDM

were killed in a car accident in this place. 28.1.2005'.[7] The lower part is written in Russian: 'Ivannov Vitali, Natalya. Killed 28.1.2005. You are forever with us. [We] remember. [We] love. [Your] children, brothers, parents'. These texts personify the dead in a number of ways: we learn who has been killed and that in this case the fatalities were two – a married couple with children; we learn that they were of Russian origin (note the code switch and the different tones in use in the different languages), and we learn of family ties and emotions working within the system of automobility (Noy forthcoming). Admittedly, emotions in automobility are not simply represented on the RDM; these memorials are an embodiment of emotions. According to the interviews, many of these monuments were constructed because of feelings of guilt that family relatives had felt toward those who were killed and the circumstances of their death.

---

7   While oftentimes the deictics 'here' or 'this' are used, automobility requires that RDMs will not be placed at the *exact* location of the car crash, which is on the road itself. Like the transposition made by documents, RDM is also located off the actual asphalt. I wish to thank Alina Liberte for her assistance in translating the text from Russian.

Figure 4.4   Ben-Simhon's RDM

In Figure 4.4 the juxtaposing of the RDM discourse with automobility is accomplished via iconicity (over and above its physical proximity to the road). While the text simply reveals the name of the deceased: 'To the memory of Ilan Ben-Simhon', the monument typically includes icons – in this case two similar steel icons of a tow truck – suggesting that the departed was a truck driver whose work was related to automobility. The only difference between the icons is that in the lower icon the personal name (Ilan) is inscribed diagonally on top of the towing track, as if being towed by it.

In both cases texts and icons reassemble and (re)claim the automobile actors – cars and people – prior to their breakdown. Hence these and other RDMs inhabit a point in the grid of automobile semiotics. In 16 (of 58) documented RDMs, authentic parts of the destroyed vehicles (cars, motorcycles and bicycles) were presented, usually welded into the stone. A multimodal type of document is produced, where the spatial location of the immobile text, together with material components of cars, challenge the official reports and suggest an alternative (im)mobilization and re-organization of the actors in the scene of the collision. If collision documents reconstruct the driver-car pair, mobilize it and (thus) stabilize it *away* from the scene of the road, then RDMs do much of the same but insist on immobilization in proximity (i.e. within visibility) to the road.[8]

The process of documenting car collisions by experts (and laymen) relies on various (auto)mobilities. Physically, this is accomplished by the speedy arrival of various agents of automobility at the scene of the collision, such as police forces, First Aid and fire-force vehicles, representatives from insurance agencies, and, when there are fatalities, also a team of the Jewish ZAKA organization, which is in charge in Israel of collecting human remains 'in order to ensure a proper Jewish burial'.[9] So immediately upon receiving a report of a car collision, which creates an instance of immobility in the heart of automobility, a whole network of communications and automobilities is ignited, intended to remove and restore parts of vehicles and humans involved in the crash.

In a case I witnessed some time ago, a collision ignited a rather different system of automobilities. This occurred on August 23rd, 2008, while I was driving in my neighbourhood in (West) Jerusalem, around 6.30p.m. As I approached the traffic light, I observed that the car that was driving ahead of me and in the same direction tried to avoid stopping and waiting at the red light by bypassing the traffic junction. This illegal and dangerous manoeuvre includes taking a right at the traffic junction when the light is red (in the Israeli traffic system, vehicles in the lane turning right are usually not required to observe the light), doing a U-turn on that road to return to the junction and then turning (right) back onto the original road beyond the traffic light. The vehicle was a tow truck, the kind on which private automobiles are lifted and transported. But when the tow truck speedily crossed the lanes of the perpendicular road it hit a scooter that was driving there (a pizza delivery scooter).

As a result of the severe blow, the young scooterist lay on the road for a few seconds, unmoving. Having seen all of this happen, I immediately called the police and the First Aid emergency lines, reporting a car accident with an injured person. Also, I used my mobile telephone to take pictures of the vehicles involved. The police and the First Aid took their time, and after a few minutes had elapsed I called both agencies again. But I was not the only one making hurried phone calls from the scene of the collision. The young man who drove the assailant vehicle also made phone calls to friends of his, who were apparently also in the business of automobility. They arrived at the scene very quickly (there is a large industrial area with many car mechanics and garages nearby), and immediately started 'fixing' it. First, they encouraged the scooter driver to get on his feet, and to move off the

---

8   In a number of interviews with family relatives associated with the construction of RDMs, they indicated that they left their telephone number on the monument in case road workers or planners would wish in the future to make changes in the road. Hence, while monuments are physically and symbolically immobile, the dialectic of im/mobilities remain.

9   Quotation from www.zaka.us (accessed: January 27th, 2009). ZAKA is an Ultra Orthodox Jewish-Israeli organization, which has become highly visible on Israeli roads.

road and sit on a nearby pedestrian bench. At the same time, they started fixing the
scooter and reassembling its broken parts. Within a few minutes, one could hardly
tell that a serious collision had taken place. The police arrived a few minutes later.
The officer (who was not in uniform and was with his family), did not leave the
police van and instead of inquiring with the injured scooterist (still confused and
shocked), had a brief exchange with those who tampered with the evidence. The
ambulance too was late and the paramedic in charge insensitively insisted that the
scooterist first admit that he wanted First Aid treatment (this is required by law).
As the young driver hesitated (later telling me that he was not sure he wanted his
workplace to know of the collision), the First Aid team left immediately without
even examining his injuries.

The depressing emotions and helplessness that this episode raised were due
to the fact that the worse part was not the preventable collision itself, but the
criminal way it was covered up, the lack of minimal civic collegiality on behalf
of the perpetrators, and the outrageous negligence evinced by the authorities.
That aside, the point of the story is that car collisions ignite different types of
(auto)mobilities, which share the task of *erasing the collision, its remains
and consequences, and reconstructing the actors in a way that satisfies the
needs and approaches of the powers that be.* This move of reconstruction can be
accomplished by official agencies or, as with the collision I witnessed, by informal
actors who can effectively operate within automobility. In light of Beckmann's
(2004) type of analysis, in this case too the witness(es) – myself and others –
stopped their cars and, moreover, disembarked from them. This means that the
collision did not sever only the scooterist and the scooter, but also disrupted other
actors in their scene from their vehicle, thus ad hoc (re)configuring the relationship
between various vehicles and their drivers within automobility.

Conclusions: towards the *motile-text hybrid*

From a semiotic perspective, discourse and mobility are interlinked and are
mutually informative: discourse charges mobility with meanings and vice versa,
mobility charges discourse with meanings. What eventually performs meaning in
the social realm is neither one (when taken separately), but a gestalt combination of
both, which can be called the *(im)mobile-text hybrid*. Since the concept of motility
nicely captures the dialectics and complex (paradoxical) interrelations between
mobility and immobility, the *(im)mobile-text hybrid* can also be called the *motile-
text hybrid*. Reciprocal semiotics, which underlie the emergence of the motile-text
hybrid, can be studied only when methodological sensitivities and epistemological
sensibilities are practiced in relations to both discourse and (im)mobility.

First and foremost, research must resist the temptation to analyse discourse
outside the ecological environment wherein it functions performatively. As the case
studies illustrate, there is never an authentically abstract or unmoving discourse.
If the text is abstracted from its situated, mobile performance – for instance, by

neglecting to account for symbolic and material values of the private automobile – the semiotic picture remains crucially incomplete. In the present case, this abstraction concerns the fact that insofar as BSs are material objects, they come to be socially meaningful as they are physically attached to other objects which are, in one way or another, part of the system of automobility. By succumbing to the temptation to analyse texts of sorts by their representational value, texts are actually decontextualized, which, in terms used in this chapter, means that they are moved (transposed), by the authority of the researcher. This in itself is not necessarily an error (cf. Latour 1987). However, this means that reflexivity is required in order to trace the translations (in the Latourian sense), transpositions and mobilities that the researcher her- or him-self has endowed.

This is true for exploring mobility as well. A mobility sensitive perspective suggests a qualitatively different type of appreciation than that which traditional research would arrive at. There are various terms through which these differences can be discussed, but basically, the mobility paradigm and within it the field of automobility is interlinked to other cutting edge developments. These include material culture and studies in society, science and technology. These approaches reject traditional (structural) focus on systems of representation. Mobility studies focuses less on *what* this or that means, and more on how the textual object of the document moves, in what circles of distribution, in what trajectories of power, to what aims, with what authority, and what types on mobilities and immobilities are at play as it travels. In the accelerated world of late-modernity, these qualities are formative in terms of creating meaning. The qualitatively different type of knowledge that these sensibilities and sensitivities produce is evinced in the different approaches to BSs and RDMs employed above. Recent publications dealing with these topics, some of which have the title 'mobile discourse', miss the point because their appreciation of mobility is limited to a metaphoric allusion, which is usually restricted to only one mode of mobility. Of my own research experience I have learned how different findings and discussions emerge when more or less the same objects or texts – or textual objects – are under examination from different approaches.

The motile-text hybrid proposes that mobilities are discursivized and socialized, and for this reason they are also multiplied. After all, accounting for the materiality of the motile-text hybrid is only the first stage. That fact that these textual objects are discursive multiplies their mobilities and brings into their semiotic performance imagined, emotional, intertextual and other mobilities. In fact, in many ways texts make the mobile object socially hearable and visible. Note that this touches on the paradox of proximity of the motile-text hybrid in automobility: RDMs and BSs invite viewers to read their inscriptions, which can be accomplished only when very close. And yet in both cases this type of proximity is downright unsafe.

Lastly, in a recent critical appraisal of *The Practice of Everyday Life*, Thrift (2004) rejects de Certeau's (1984) strong discursive proposition, yet he suggests that other types of discourse are infiltrating automobility through the many high-tech components (and their software texts) that are nowadays installed in cars. While this perspective is promising, this chapter attended to low-tech types of auto(im)mobile texts. The proposition is that regardless of whether the discourse is located on car rear windows or in CPUs under the hood, or whether it is high-tech or low-tech, as the research and theorizing of automobility expands we acknowledge the enmeshment of materiality, discourse and (auto)mobility in establishing meanings in situated ways, embodied in the *motile-text hybrid*.

References

Appadurai, A. 1986. *The Social Life of Things: Commodities in Cultural Perspective*. Cambridge: Cambridge University Press.

Beckmann, J. 2004. Mobility and safety. *Theory, Culture and Society*, 21(4–5), 81–100.

Bloch, L.R. 2000a. Rhetoric on the roads of Israel: the assassination and political bumper stickers, in *The Assassination of Yitzhak Rabin*, edited by Y. Peri. Stanford, CA: Stanford University Press, 257–79.

Bloch, L.R. 2000b. Setting the public sphere in motion: the rhetoric of political bumper stickers in Israel. *Political Communication*, 17(4), 433–56.

Böhm, S., Jones, C., Land, C. and Paterson, M. (eds) 2006. *Against Automobility*. Malden, MA: Blackwell Publishing.

Brottman, M. (ed.) 2002. *Car Crash Culture*. New York: Palgrave.

Certeau, M.d. 1984. *The Practice of Everyday Life* (translation by S. Rendall). Berkeley: University of California Press.

Clark, J. and Franzmann, M. 2006. Authority from grief, presence and place in the making of roadside memorials. *Death Studies*, 30(6), 579–99.

Featherstone, M. 2004. Automobilities: an introduction. *Theory, Culture and Society*, 21(4–5), 1–24.

Featherstone, M., Thrift, N.J. and Urry, J. (eds) 2005. *Automobilities*. London: Sage.

Hamadi, I.B. and Chittajallu, S. 2008. *IRF – World Road Statistics 2008*. Geneva, Switzerland: International Road Federation.

Hartig, K.V. and Dunn, K.M. 2002. Roadside: interpreting new deathscapes in Newcastle, New South Wales. *Australian Geographical Studies*, 36(1), 5–20.

Henzel, C. 1991. Cruces in roadside landscape of Northeastern Mexico. *Journal of Cultural Geography*, 11(2), 93–106.

Hull, M.S. 2003. The file: agency, authority, and autography in an Islamabad bureaucracy. *Language and Communication*, 23, 287–314.

Latour, B. 1987. *Science in Action: How to Follow Scientists and Engineers through Society*. Cambridge, MA: Harvard University Press.

Livnat, Z. and Shlesinger, Y. 2002. 'Street fighting': the rhetoric of bumper stickers in Israel. *SCRIPT: Literacy Research, Theory and Practice*, 5–6, 59–80.

Noy, C. 2007. Sampling knowledge: the hermeneutics of snowball sampling in qualitative research. *International Journal of Social Research Methodology*, 11(4), 327–44.

Noy, C. 2008a. Mediation materialized: the semiotics of a visitor book at an Israeli commemoration site. *Critical Studies in Media Communication*, 25(2), 175–95.

Noy, C. 2008b. Writing ideology: hybrid symbols in a commemorative visitor book in Israel. *Journal of Linguistic Anthropology*, 18(1), 62–81.

Noy, C. 2009. On driving a car and being a family: a reflexive autoethnography, in *Material Culture and Technology in Everyday Life: Ethnographic Approaches*, edited by P. Vannini. New York: Peter Lang Publishing, 101–13.

Noy, C. Forthcoming. Inhabiting the family-car: children-passengers and parents-drivers on the school run. *Semiotica*.

Reid, J.K. and Reid, C.L. 2001. A cross marks the spot: a study of roadside death memorials in Texas and in Oklahoma. *Death Studies*, 25(4), 341–56.

Salamon, H. 2001. Political bumper stickers in contemporary Israel: folklore as an emotional battlefield. *Jerusalem Studies in Jewish Folklore*, XXI, 113–44.

Silverstein, M. 1996. The secret life of texts, in *Natural Histories of Discourse*, edited by M. Silverstein and G. Urban. Chicago: University of Chicago Press, 81–105.

Thrift, N.J. 1996. *Spatial Formations*. London: Sage.

Thrift, N.J. 2004. Driving in the city. *Theory, Culture and Society*, 21(4–5), 41–59.

Urry, J. 1990. *The Tourist Gaze: Leisure and Travel in Contemporary Societies*. London: Sage Publications.

Urry, J. 2002. Mobility and proximity. *Sociology*, 36(2), 255–74.

Urry, J. 2004. The 'system' of automobility. *Theory, Culture and Society*, 21(4–5), 25–39.

Vardi, I. 2006. *Death on the Margins: Roadside Memorials for Traffic Accident Victims in Israel and the Struggle for a Place in Collective Memory and Consciousness*. Unpublished M.A. Thesis, Tel-Aviv University, Tel-Aviv.

Chapter 5
Mobility After War: Re-negotiating Belonging in Jaffna, Sri Lanka

Eva Gerharz

We were so happy to see each other after a very long time but somehow we felt scared of them. They were very Westernized. The way they spoke was strange for us. And they were wearing nice and new dresses. We felt odd compared to them, like street people.

This is the comment of a Tamil woman in Sri Lanka's northern peninsula of Jaffna who met her relatives living in Europe after a Ceasefire Agreement had temporarily ended the Sri Lankan civil war in 2002.[1] The statement highlights the ambivalence of the reunion after more than 10 years, because it reveals her subjective feelings of alienation and that of other locals during the war vis-à-vis her emigrated relatives who had temporarily returned for a visit. Despite all the joy and happiness which marked the moment of reunion, this quote hints at the material and emotional distance which had emerged within a family and, more generally among the Sri Lankan Tamils, quite often constructed as a unified and homogeneous, yet spatially dispersed community under the conditions of war-related immobility and distance.

Taking this observation as a starting point, the article addresses the new opportunities for mobility after a long period of immobility. My aim is to explore, how social relationships change when migrants travel back 'home'

---

1 The Ceasefire Agreement which was negotiated between the Government of Sri Lanka and the Liberation Tigers of Tamil Eelam in 2002 was the most promising initiative for consolidating a sustainable solution to the decade-old conflict. But despite initial enthusiasm among Sri Lankans, political stakeholders and donors the peace process gradually broke down. In January 2008 the Sri Lankan government officially annulled the Ceasefire Agreement. Military offensives with an unrivalled intensity took place, which have claimed several thousand lives. Despite an international outcry incriminating the ruthless proceeding of the Sri Lankan Army, none of the institutions which could act as a peace-maker have been able to undertake any effective measures to prevent further escalation. The Sri Lankan government justified its harsh moves with their attempt to destroy the LTTE completely. In May 2009, the troops finally managed to capture the last LTTE stronghold. Almost the entire leadership and thousands of civilians died. In February 2010 large numbers of civilians are still been kept in internment-camps.

after having been absent for many years. I will show that these changes concern not only family relations, but also ethnic relations in the particular context of transnational solidarity. This results in reconstructions of ethnic belonging, in a process of re-positioning and re-adjusting belonging which is embedded into the dynamics of (im)mobility. I argue that a growing awareness of difference in experiencing cultural practices, expressions and perspectives in everyday life has emerged as a result of renewed mobility. At a more general level, the dynamics of boundary-drawing are interrelated with general patterns of globalization and the restructuration of society under globalized, yet mobile conditions. This restructuration is characterized, as Meyer and Geschiere have called it, by 'dialectics of flow and closure' (Meyer and Geschiere 1999) in which re-constructions of identity are pertinent in translocal space. The new frictions and cleavages between spatially dispersed groups result from opportunities provided by mobility and refer to life-worldly experiences embedded into social relationships within 'actual reach' (Schutz and Luckmann 1979). Social relationships within actual reach are tied to the condition of co-presence, because they presuppose face-to-face contact. Seen against the background that contemporary society is constituted by mobile communication and virtual co-presence, it is necessary to explore the elements which turn face-to-face proximity into a unique social experience entailing a particular quality. In the context of (im)mobility, I attempt to investigate the re-structuring of social relations in what has been commonly conceptualized as 'transnational community'. I will show how mobility, structured by time and space, reinforces the emergence of multiple forms of belonging. Especially in the context of migration and diasporization, but also at a more general level through the anticipation of the world as 'attainable reach' (Schutz and Luckmann 1979, see also Engelbrekt in this book), the question of 'where and when to belong' becomes particularly relevant. Exploring the notion of belonging in relation to (im)mobility reveals the enduring significance of localized co-presence in times shaped by mobile options and globalization.

Mobility, immobility and globalization

Mobility is a constituting phenomenon of contemporary society. Moving has become a normal part of everyday life for large parts of the world's population. People move on foot, by bike, on wheels, on ships or by plane, individually or collectively, for short or long periods of time. The motives and motivations vary greatly, as does the scale of movement. People travel for business, as tourists, to meet with family members or friends, to work or to save their lives from environmental hazards, war and other disasters. The practices of mobility, overriding geographical space and travel, are related to imagination and anticipation of distant places and locales. If we did not know about the existence of distant places, and if we did not have personal relationships to people there, most of us would remain in the localities where we spend our everyday life.

Mobility, as an analytical concept, is related to attempts to understand modern society shaped by and shaping globalization. A large body of literature has discussed the diverse phenomena characterizing the process of globalization as well as attempts to understand the state of the 'global situation' (Tsing 2002) under conditions of modernity. A guiding principle in understanding the conditions of the globalized society is what Harvey (1989) and others describe as time-space compression. The observation that new transport and communication has subdued space and compressed time forms the starting point of most globalization research. Especially the significance of geographical space has been intensely discussed. The often neglected aspect of time has been highlighted by Bauman (2000) who describes the consequences of everything moving faster and shows how this affects spatial reordering in 'Liquid Modernity', a term which highlights deterritorialization or detachment. The changes accompanying global mobility have led some scholars to proclaim that territoriality like localities will be increasingly insignificant. However, this has instigated manifold attempts to prove the enduring significance of the local. Robertson (1995) who argued that homogenizing and heterogenizing effects lead to 'glocalization' was among those who ensured that localization as a crucial process structuring globalized society is taken for granted in today's scholarship.

Mobilities are central to the re-ordering of space and time instigated by globalization, because they transcend both. Theoretical approaches which relate space-time compression to globalized society, reveal that, first, various forms of mobility cause places to move closer together, and that every form of movement speeds up. Secondly, global flows affect the reordering of social realities and expressions of cultural difference in localities and places. At the same time, people keep on moving, represent, and reproduce global flows. This has spurred scholars to elaborate on the significance of globalizing processes for culture and identity. Some have argued that global flows and cultural contact produce hybridizations (Nederveen Pieterse 1995) or creolization (Hannerz 1987), describing the commingling of different cultural contents. Bringing together identity and belonging with space and mobility, Appadurai has introduced the concept of ethnoscapes to describe the,

> landscape of persons who constitute the shifting world in which we live: tourists, immigrants, refugees, exiles, guest workers, and other moving groups and individuals constitute an essential feature of the world ... This is not to say that there are no relatively stable communities and networks of kinship, friendship, work, and leisure, as well as of birth, residence, and other filial forms. But it is to say that the warp of these stabilities is everywhere shot through with the woof of human motion, as more persons and groups deal with the realities of having to move or the fantasies of wanting to move. (1996: 33–4)

Mobile people constituting ethnoscapes establish translocal (Lachenmann 2008) or transnational spaces (Pries 1999, Faist 2003) or fields (Basch, Glick Schiller and Szanton Blanc 1994). This reveals a 'complex interrelation between travel and dwelling, home and not-home' (Hannam, Sheller and Urry 2006: 10). Migrants move to other places and at the same time, they maintain contact with their place of origin. Translocal spaces emerge,[2] which do not necessarily transcend national borders but which stretch out in geographical space.

Mobility is about transcending geographical space, and more than that. Concepts of mobility are quite extensive and comprise different forms of physical movement of people, collectively and individually, over long and short distances, virtual mobility and social mobility (see Urry 2007). Limiting ourselves to geographical mobility, Urry's way of distinguishing between different forms of travel provides a useful framework for analysing the modes of mobility which are relevant in the empirical case. 'Corporeal travel' describes human movement and migration. Likewise, the 'travel of objects' constitutes a meaningful practice of mobility. The use of communication technologies can be understood as either 'imaginative travel' when images move across print and visual media, as 'virtual travel' in the sense of transcending distance in real time, or as 'communicative travel' through person-to-person messages (Urry 2007: 47). Transnational or translocal spaces are made up by social relationships spanning geographical distances, which do not necessarily need to rely on physical mobility any longer. Communication technologies such as the telephone and the Internet are constitutive of what Castells called the 'Network Society' (2000) and subdue space and time. One might argue that advanced communication technologies might compensate for a lack of 'physical mobility' to a certain extent. The case investigated in this article will show whether this assumption is supported by empirical evidence.

When we speak about mobility, we also have to consider immobility. Hannam, Sheller and Urry (2006) argue that mobility can only be understood in the function of its oppositional constituency that is immobility. Borrowing Harvey's notion of the 'spatial fix', they state that 'infrastructural and institutional moorings' (2006: 3) necessarily configure and enable mobilities. This means that mobility always entails a relational dimension: without immobility, there is no mobility. Some scholars working on mobility conceptualize immobility mainly in the sense of moorings resembling the immobile knots of a mobile network, so-called platforms such as transmitters, roads, garages, stations, airports etc. (Urry 2007: 53, Hannam, Sheller and Urry 2006; Adey 2006). Immobility, however, can also relate to the symbolic construction of locality in translocal space. People constantly localize themselves, even when they are on the move. Exiles, diasporas and labour migrants try to make a home wherever they reside. Conceptualizing locality as a construct of phenomenological quality (Appadurai 1996, Pfaff-Czarnecka 2005), the local appears as a space where everyday life takes place and is made up of face-to-face

---

2   I use the term 'translocal space' to highlight the relational and multi-dimensional character of transcending space. In contrast to the notion of 'transnational space', which relates mainly to the crossing of national boundaries, translocal space stresses not only geographical mobility but relates to symbolic and imagined boundaries in space and time.

contact which entails a special kind of sensual experience (Hannerz 1996: 26–7). The case of northern Sri Lanka to which I am referring to in this article draws attention to this kind of immobility as dwelling in locality. Hannam, Sheller and Urry (2006: 3) have argued that mobilities are structured by power hierarchies and that the rights to travel, for example, are highly uneven and structured by gender, class, status, and ethnicity. In my empirical example immobility emerges as enforced immobility, which is imposed by the state in the course of civil war. This enforced immobility is related to different forms of mobility (imaginative, virtual, communicative, and corporeal) which is structured by temporality: while mobility was restricted during civil war, the peace process ushered in a period of manifold mobilities which produced new configurations of belonging.

Jaffna Tamils between mobility and immobility

Shortly before my first trip to Sri Lanka in 2002,[3] a Ceasefire Agreement ending two decades of the civil war, which had begun in the early 1980s, was signed between the Sri Lankan Government and the Liberation Tigers of Tamil Eelam (LTTE). Although this ceasefire lasted only until 2007, it was a historical event for many and set loose a whole array of changes especially in the north-eastern territories, where the majority of the ethnic Tamil minority lives. It is also the north-eastern parts of Sri Lanka where the LTTE, which attempted to represent Tamil claims for independence, had its stronghold. The many years of intense fighting had turned the north-east into a war zone. Until the Ceasefire, the war had caused large-scale destructions and more than 67,000 deaths, and huge numbers of Internally Displaced People (IDP), refugee movements to neighbouring India and other countries. Especially the Tamil-inhabited peninsula of Jaffna located in the north, which is regarded as the centre of Tamil culture as well as resistance, has been terribly destroyed by the fighting erupting again and again. Before 1995, Jaffna was under the control of the LTTE. A military offensive brought the peninsula under government control which has prevailed until today. But only the Ceasefire led to a re-opening of this area, which remained cut-off from the mainland throughout the war. This was accompanied by large-scale reconstruction and development efforts and intensified circulation of goods and people.

It was also in Jaffna that re-migrating and circulating migrants were most visible. Rough estimations by local authorities reveal that the population has been halved since the outbreak of war. Concerning the population living in the diaspora, it is

---

3   This chapter is based on the findings of my ethnographic fieldwork which I conducted in Sri Lanka between 2002 and 2004. I twice spent six months in different parts of Sri Lanka, but mainly in the northern peninsula of Jaffna. The methodological tools employed are comprised of participant semi-structured and open unstructured interviews with various stakeholders, observations in various fields, analysis of documents and other materials (see Gerharz 2007).

estimated that 90% of Sri Lankan Tamils living abroad are from Jaffna (Gunaratna 1998). Although the war caused massive movements of people, migration from Jaffna has a long history dating back to (post)colonial times. Compared to Tamils from other parts of Sri Lanka the Jaffna Tamils were comparably well-off and aspirations for education had produced (temporary) migratory movements to Western countries. In addition, the shortage of appropriate jobs forced Jaffna Tamils to migrate to southern Sri Lanka and other British colonies (Cheran 2001). The rise of anti-Tamil sentiments and discrimination also reinforced migration after independence. When the war broke out in 1983 many Tamil had already found a new home in Canada, the US, Australia, Europe and other countries. Others who were planning to return after completion of their education decided to stay abroad. As the war escalated, huge numbers of refugees fled to India, whereas movement to Western countries continued. The chance to migrate depended on people's individual resources and was structured by caste, gender and other determinants.

Group formation has taken place among the diaspora[4] in many ways. This includes social and welfare associations assisting diaspora members, cultural organizations, media, business and trade, and political organizations (Cheran 2001, 2007, McDowell 1996, Fuglerud 2001). These social organizations have two major functions. On the one hand, they serve the needs and group formation aspirations of the diaspora itself. There are, for example, organizations offering courses in Tamil language, dance, or music for second generation migrants or those assisting elderly Tamil migrants. On the other hand, there are organizations whose work is mainly directed towards the homeland and consists of lobbying, public demonstrations and information exchange. The structures mobilizing for diasporic identity formation and political lobbying not only include informal and formal forms (Wayland 2004: 415), but are also closely interlinked. Many organizations fulfil both functions and relate cultural work in the diaspora to activism for the homeland. Cultural shows, for example, are often organized with the aim of raising funds for relief and development, but also for war-funding. Political and non-political organizational forms often overlap and their objectives are fuzzy. The same relates to businesses in the diaspora, of which many are allegedly run by the LTTE, which oversees a strong transnational network. Group formation in the diaspora thus constitutes an important prerequisite for the construction of transnational space. The mechanisms to preserve Tamil identity are related to a strong homeland orientation. This constitutive aspect of 'diaspora'

---

4   I use the notion of 'diaspora' in the context of Sri Lankan migrants, for two reasons. First, I pragmatically adopt the term many Tamils living abroad use themselves, and which has also become mainstream in the relevant literature. Secondly, I argue for the usage of the diaspora concept by referring to Brubaker's (2005) account which gives a useful and encompassing overview of this vast debate. He claims that dispersion, boundary maintenance vis-à-vis the host society and a homeland orientation make a diaspora. The latter includes a collective memory, an eventual will to return, commitment to its maintenance or restoration and a sense of belongingness and identification with the homeland (Brubaker 2005: 5).

(Brubaker 2005) contributes to the formation of transnational social spaces consisting of migrants' practices in maintaining, building and reinforcing links with the places of origin as well as with other places where diaspora members are located. Diaspora Tamils have developed different strategies to establish these spaces ranging from media and performative representations to the establishment of co-presence through travel.

Different forms of mobility can be distinguished as relevant. Following Urry (2007: 47) the first form of mobility can be described as 'corporeal travel', describing human movement across space in the form of migration and travel. Tamils migrated in large numbers yet under harsh conditions to seek refuge abroad, in neighbouring India or on other parts of the island. But in contrast to other contexts in which homeland visits have been a powerful instrument for maintaining transnational social spaces for a long time (Scheyvens 2007, Bruner 1996), Tamils from Northern Sri Lanka have had limited possibilities to return. Many parts of the northern war zones remained inaccessible by land for military reasons. Travel to and out of Jaffna was only possible by air or sea, and was limited by harsh restrictions and insecurity. In Jaffna, people relied on dangerous travel routes to the south and many lost their lives (Suryakumaran 2002). The second form of mobility is 'imaginative travel' affected through the images of places and people, which appear in print and visual media.[5] Among Jaffna's locals and migrants, these possibilities were severely limited in both directions. A road block and unreliable postal services prevented the transportation of print media to and from Jaffna. Getting images from Jaffna to the outside world was difficult because of the absence of electronic media and because journalists were not allowed to report from the war zones. Electricity and telephone services were suspended most of the time. This may also be among the reasons why the maintenance of a sense of a collective stronghold related to the homeland and group-cohesion was so important: in order to get information about those who stayed behind, it was essential to establish and remain in contact with newly arrived migrants. This leads to 'communicative travel' as a third form which appears as highly relevant in this context. Because mail service and contact by telephone were unreliable, Tamils became creative in developing strategies to spread information. One of my interviewees said that before the Internet, people used to call a number in London, where an answering machine reported about the newest developments in the homeland.

Discussing the different forms of travel highlights the importance of immobility as the counterpart of mobility. Modes of mobility were severely restricted and limited before 2002. Restrictions were imposed on corporeal travel and circulation of migrants and foreigners, including journalists, correspondents and aid workers. Information flows from Jaffna through imaginative or communicative travel were

---

5   Imaginative travel resembles to a certain extent what Appadurai called mediascapes (Appadurai 1996: 33). But while Appadurai refers to the global scope of imaginative flows, Urry's concept allows us to focus more on connections and flows between certain locales.

scarce and depended on face-to-face communication. In contrast to moorings as connecting points of mobilities (Urry 2007: 54, Hannam, Sheller and Urry 2006, Adey 2006), immobility appears as an empirical phenomenon materialized in terms of the restricted movement of people, goods, images, information and knowledge imposed by the state.

When the Ceasefire was signed and the road to Jaffna was reopened in April 2002, the relationship between mobility and immobility changed considerably. This 'opening to the world' (Gerharz 2008) was accompanied by a massive movement of people and goods. One university professor in Jaffna whom I interviewed in 2003 described the consequences of this process illustratively: 'Before the ceasefire, Jaffna was a closed prison. Now it has become an open marketplace'. This indicates that local everyday life was disrupted by the new forms of mobility and their intensity impinged on the local in different ways. Jaffna was re-integrated into global and national economic exchange networks. Consumer goods, which had never been available in Jaffna before, produced changes in the economic sector by opening new markets and niches and remarkably transformed consumption patterns. All of a sudden, exotic goods such as Swiss chocolate and American chewing gums were available in newly constructed supermarkets, to which people first had to get used. Jaffna also regained connection to global mediascapes (Appadurai 1996: 33). The availability of satellite TV changed local media consumption. Especially Tamil channels, broadcasting from neighbouring Tamil Nadu in India, attracted much local attention and were watched extensively. The market for Indian and Western DVDs boomed. Newspapers issued in Colombo were available in the local market the same day and even Internet cafes mushroomed. Free movement produced 'tourists' from southern Sri Lanka, India, and Western countries who wanted to see the peninsula. They, along with international staff employed by the various development organizations, visibly represented cultural difference. All these changes and new images Jaffna became exposed to contributed to a translocalization of people's life-worlds. The atmosphere I experienced among Jaffna residents was determined by enthusiasm about free movement and openness, but, at the same time, by ambivalent reactions to new images and impressions. The new forms of knowledge, which became available after the Ceasefire, changed perspectives on the world, but also on the local space itself and the 'we-group' (Elwert 2002) living therein and beyond.

Enthusiasm and alienation: encounters under conditions of mobility

Jaffna's exposure to global mediascapes changed local perspectives in different ways. Thanks to communication facilities allowing imaginative and virtual travel (satellite television, up-to-date newspapers, radio and Internet) people in Jaffna were more aware of what was going on in other parts of the world and reinterpreted global events with regard to their own situation and their effects on the local context. The imaginative and virtual travel via television alerted many to the fallacies of a Westernized life-style, which influenced their interpretations of foreigners' actions and behaviour in Jaffna itself. Such immoralities, however, impinged on the local in the sense that it invited imitation. In the newly created Internet-cafes I witnessed many young men looking at dingy websites called tamilsex.com, hindisex.com and so forth. For those locals recognizing these new popular leisure pursuits of their youngsters, this was quite often related to the impact of the alien and 'spoiled' practices which had found their way into Jaffna after the re-opening.

Local perspectives changed also as a consequence of the physical travel of development experts, tourists, diaspora members and Jaffna Tamils. Cultural contact resulting from this kind of mobility led to the emergence of a translocal interface, where all kinds of differences were negotiated. Based on the ethnographic research perspective applied, I discovered the changes emerging from the re-established contact between locals and circulating diaspora Tamils. Many visiting Tamils reported very ambivalent experiences. I was sometimes approached by migrant Tamils on Jaffna's street who wanted to exchange information about how I was experiencing post-war Jaffna and many were eager to express their emotional involvement as well as alienation. Some of them differed a lot from the locals and a few looked much more the 'tourist' than I myself felt. Having lunch in one of diaspora Tamils' favourite hang outs, the Palm Beach Restaurant, I was stunned when a family entered, all of them well fed and dressed in shorts. The father was wearing a baseball cap and had a camera dangling around his neck. Like some diaspora Tamils, these Australian Tamils expressed alienation from what they had self-evidently considered their homeland. The daughter who worked as a nurse and who had never been to Jaffna before complained forcefully about the local hygiene: 'This place is so filthy! I can't use the toilets here, it's so terrible'. Other diaspora Tamils I spoke to explained that they could not imagine migrating back to Jaffna in the near future. They referred to inadequate medical care and the deficient school system when explaining that it was impossible to live there with small children. Other diaspora Tamils reported about having had alienating experiences although they tried to adapt to local circumstances. One young Tamil who had spent most of his life in Great Britain and who volunteered in a local NGO for one year claimed:

> I feel localized in everyday life and during work. But people see me as a foreigner. I speak Tamil, but they can differentiate between my accent and the local one. I try to speak like them, but nevertheless, they recognize it. Moreover, they don't know who I am, who my parents are and where I live. Then there are the many little things. For example, I am used to carrying a backpack, because it is more practical. But the fact that I am carrying a backpack, makes me a foreigner.

Full of ambivalent impressions, this quote hints at the construction of difference, which is embedded into a set of hierarchy and power relations. This also includes economic differences, which were highlighted by the woman quoted in the beginning of this chapter. Like the young Tamil from Great Britain, she referred to the way her relatives spoke to her and her family and the language they used, though they spoke Tamil, and the way they were dressed. The way she highlighted the differences in appearance pointed to a hierarchical relationship between Jaffna and the West. For her, the dress symbolized existing economic differences between the developed West and Jaffna. This caused feelings of being inferior and made her recognize economic difference as a dividing factor between her family and the relatives from Europe. Others took such observations with a sense of humour and laughed about a man who had returned from Malaysia and had become fat there.

Constructions of differences between Jaffna and the outside world, particularly the West, were linked to cultural identity as well. In local images, the 'culture' of Jaffna people was constructed as morally superior to others. The fear that the local culture is being spoiled by Western influences represented by visiting diaspora Tamils was prevalent in many discussions. Such difficulties arose mainly between generations and put a strain on family ties. In one case I was told about a Tamil grandmother throwing her visiting granddaughters out of the house and calling them prostitutes. According to my informant, the girls had not done anything wrong except not observe the local dress code. For the grandmother, wearing Western clothes symbolized the girls' immorality. Similarly, some people assumed that 'something must go wrong' when Tamil girls spent their leisure time at the local beach in swimsuits. It is often assumed that girls and women are particularly threatened by foreign culture and that they might become morally spoiled or loose, especially in migration contexts (Dannecker 2005: 660). As much as in other parts in South Asia, dressing is central in Jaffna because wearing immodest clothes is related to sexual promiscuity. But also images of how to behave in the public arena are gendered in a different way and produced conflict between local and diaspora Tamils. A Catholic priest told me about a grandmother who complained about her granddaughter: 'How can this girl go in a car alone? She must be someone else's grandchild!' The construction of a causal relationship between Western clothing and immodest behaviour in public also caused prejudices and exaggerated images of sexual behaviour. One hotly debated issue, relevant also in the developmental context, was HIV. This disease was unknown in Jaffna before the Ceasefire. Only after the re-opening approximately 30 cases became known, probably because testing methods were introduced by Western development organizations shortly after the Ceasefire. When I discussed the occurrence of HIV cases with some local Tamils, I was told that it was caused by the reopening. Who exactly was to be held responsible remained unclear. While some people argued that international staff of development organizations brought HIV, others argued that it was spread by a returned diaspora woman.

The empirical encounters presented show that diaspora and local Tamils stressed certain differences between the local and the foreign in the course of re-established contact. Diaspora Tamils who travelled to northern Sri Lanka after many years of absence were at times alienated by the ways they experienced the local Tamils and their way of life. This related not only to constructions of economic and social disadvantages, but to social relations in a broader sense. Some diaspora Tamils complained about tight social control within families and neighbourhoods, about the local narrow-mindedness and other characteristics of Jaffna society which can also be traced back to the fact that it is shaped by the rural context to a large extent. Similarly, for some of the local Tamils diaspora members represented otherness in a critical way. Some even rejected the strange representations. These processes of reciprocal boundary-drawing and exclusion are based on images and constructions of the West in opposition to the local, the foreign versus the well known. Diasporization thus, has contributed to the emergence of de-localized, translocal spaces in their own way, reconfiguring notions of identity and belonging. However, although oscillating between processes of localization and de-localization (Spiegel 2005: 159), there is a sense of consciousness about shared experiences among diasporas and locals as many studies on 'transnational communities'[6] have revealed. In the case of the Tamil diaspora, it is the experience of diasporization and the identification with the imagined homeland which constituted the space for transnational community building. But over time, and in situations of re-established co-presence as described in this chapter, the Ceasefire Agreement marked an important turning point which led to a re-ordering of transnational social relations. That diasporization in the sense of being away from the homeland changes identity formations has been revealed by Cheran who argues:

> recent technological changes and diasporic existence have had an immense impact on perceptions of mono dimensional identities. Diaspora produces multiple identities and hybridities such as Tamil-German, Tamil-Canadian, Tamil-Norwegian and Tamil-Dutch. There is a generation of 'Tamils who may not want to be identified solely as Tamils'. (Cheran 2004: 272)

But can we assume that these hybrid identities per se relate to local identities of those remaining in the place of origin? Or aren't the patterns of identification and constructions of belonging embedded into much more complex, multi-layered processes? My empirical material challenges precisely this assumption. Dynamics of boundary-drawing between local and diaspora Tamils reveal that perceptions and constructions of commonality have changed as well. There are issues with which diaspora Tamils and local Tamils highlight their commonalities and there is a strong agreement on sharing one Tamil identity. This is expressed, for example,

---

6   Here, again, it is important to be aware of the dangers of essentialization. Similar to identity, communities are social formations which are constructed, and also instrumentalized for representation. Adamson notes in this context that interlocking networks characterize a transnational community as constituting transnational social space which is partially embedded in and interacts with other networks, institutions and social spaces. At the same time, its quality as partially autonomous space defined by a common homeland experience is retained (2002: 159).

through solidarity and united representations in the context of the struggle for an independent homeland, a point on which many (not all) diaspora and local Tamils agree. Although Tamilness entails a globalized dimension related to some universal characteristics of Tamilness (language, literature, performing arts) which are constructed across boundaries, the discrepancies highlighted above hint at more controversial dimensions and revealed difference at the level of everyday life which set limitations to the concept of identity. In the following section I seek to critically explore the usefulness of 'belonging' as a notion, which allows us to grasp the ambivalence and complexity of diaspora-local-relationships in this particular context of (im)mobility.

Multiple forms of belonging

To comprehend the different layers of collective identification and demarcation I propose to investigate the usefulness of the concept of belonging to grasp the complex nature of social positioning in translocal space. Taking the re-established direct contact and co-presence as a critical turning point, the notion of belonging diverts our attention to how collective attachments and mutual commitments change and how commonality is constructed. Pfaff-Czarnecka (2008) has argued that these three dimensions of belonging (commonality, mutuality and collectivity) determine the ways in which social relations in transnational space are constituted. Moreover, she criticizes how in many studies, commonality of interest, mutually accepted division of labour, common aspirations and expectations are taken for granted without considering inequalities and power differentials. She reveals that:

> Transmigrants and those remaining at home tend to be depicted as sharing the same goals, interests and political attitudes … But transnational space is forged by networks of diverse political allegiance. 'Local societies' do not necessarily share political ideologies. On the contrary, factionalism, i.e. political group formation cutting across socio-economic lines is often stabilised through patron-client relationships, is a wide-spread phenomenon in local societies around the globe. (Pfaff-Czarnecka 2008: 317)

Translating the discrepancies described by Pfaff-Czarnecka for political activism in transnational space into more mundane concerns brings the concept of belonging to the centre of attention. This is important if we attempt to understand and conceptualize the highly ambivalent dynamics of meeting again as it has been described for the case of Jaffna Tamils. These co-present encounters were highly emotional and determined by ambivalent feelings of sameness and difference. According to Yuval-Davis, Kannabiran and Vieten (2006: 1), the emotional dimension is central to notions of belonging. Moreover, belonging encompasses and relates both, citizenship and identity. They furthermore argue that the politics of belonging[7] can be viewed as situated in three different, yet complementary ways (Yuval-Davis, Kannabiran and Vieten 2006: 7). These are central if we want to understand the dynamics of constructing sameness and difference at multiple junctions, as well as the fragmented and situated nature of belonging. The first way of situating belonging relates to temporality. Issues of belonging are embedded into historical, technological, economic and political developments. The second dimension is that belonging is spatially situated which refers to local and regional difference, for example concerning responses to globalization. Thirdly, belonging depends on intersectionality, highlighting that different intersecting and intermeshing dimensions of belonging (class, ethnicity, race, gender, sexuality, stage of life cycle and so on) are related to each other in different ways and intensity. These three dimensions provide a constructive basis for analysing the effects of the various forms of mobility which decisively changed everyday life in Jaffna. With the concept of belonging defined thus, I seek to move beyond the analytical limitations inherent in the concept of identity, especially the 'residual elements of essentialization retained even within the idea of fragmented and multiple identities' (Anthias 2002: 491). Comprehending belonging as intersecting social positionings expressed in terms of commonality, mutuality and collectivity, as Pfaff-Czarnecka suggests, the three parameters of intersectionality, temporality, and spatiality as proposed by Yuval-Davis and her colleagues provide a fruitful and constructive tool for analysing the multiple and ambivalent forms of belonging which I observed in re-opened, post-war Jaffna.

Intersectionality

In the Sri Lankan Tamil context, intersectionality as one dimension of the dynamics of belonging is particularly relevant with regard to gender and age. I have shown above that gender constructions are crucial in the process of situating diaspora Tamils vis-à-vis local culture. For Sri Lanka especially, gender has been repeatedly employed as a constitutive element for the construction of nationalism and identity (Maunaguru 1995, Rajasingham-Senanayake 2004, Tambiah 2005). Although LTTE's policies and practices on gender and women have challenged older constructions in multiple ways, women are still constructed as the bearers of tradition and culture. Considering the centrality of these constructions for determining the level of cultural purity it can be argued that the meaning of womanhood is to a large extent equated with Tamil culture. Therefore, 'spoiling culture' in Jaffna has quite often to do with gender relations. Westernization in this context is regarded by some defenders of local tradition as dissolving the separation of 'traditional' separate spaces and questions women's sexual promiscuity. As a consequence, unmarried Tamil women who do not meet the

---

7 Yuval-Davis, Kannabiran and Vieten refer basically to the politics of belonging and less to the core concept itself. Nevertheless, their categories point at crucial dimensions which are highly relevant for the concept of belonging.

local norms of gendered behaviour are considered as 'loose'. This 'looseness' or 'being morally spoiled' is ascribed to women who do not meet the local norms for dressing or behaviour in the public arena for example when drinking alcohol, driving cars, or spending their leisure time at the beach in swimsuits. Wearing jeans instead of a Sari or other South Asian clothing can represent the potential dangers for local culture which needs to be protected against foreign, mainly Western influences. These accusations put considerable pressure particularly on younger diaspora women belonging to the second generation who travelled to Jaffna sometimes for the first time in their lives. They are constructed as an embodied danger for local culture which needs to be preserved. Re-positioning and re-constituting belonging thus becomes an important project for these women vis-à-vis local Tamils. Gendered identity intersecting with diaspora Tamilness provides an interesting example of the significance of different positions and belongings, which are embedded into time and space as well.

Temporality

Many Jaffna Tamils who travelled back to see the peninsula after having been absent for many years had particular constructions in mind, which were filled with images from the past. These nostalgic images were intensively celebrated on several occasions. As an observer I participated at one school located in Jaffna where several elderly diaspora and local Tamils belonging to the schools' alumni organization had gathered together with their old teacher. The meeting was centred on chatting about the old times when they were living as students in Jaffna. Apart from their memories, they exchanged proverbs and poems about 'Old is Gold'. At the end of their meeting, they sang a song together praising the 'Good Old Days'. Such rituals related to Jaffna's glorious past reinforced a sense of commonality between elderly men, regardless whether they had migrated or not. This shows how generational, gendered and Jaffna Tamil identities intersect and are embedded in the temporal order of social reality. The orientation towards the locality's 'status quo ante' antagonized members of the younger generation of Jaffna Tamils, whose life-experience is shaped by war, rather than the peaceful good old days. Members of the same age cohort have been raised abroad, exposed to completely different circumstances of living, whereas gendered images and positions differ as well.

    Apart from these complex frictions leading to different patterns and dynamics of belonging, temporality determines more general processes of positioning among diaspora and local Tamils. Schiffauer (2006: 175) describes the relationship between migrants and those who stayed behind as particularly tense in the context of civil wars. The question 'what would have happened if I had stayed back like others' is among the most depressing misgivings among migrants. Likewise, Schiffauer argues, refugees are sometimes labelled as traitors who did not contribute to defending the homeland. During fieldwork, I was confronted with subjective and collective feelings of guilt among migrant Tamils, but I hardly ever heard that they were accused of having escaped. Nevertheless, different life-worldly experiences in the diaspora and at home result in a complex frame of reference for both, which are embedded into a peculiar relationship between sameness and difference. At this point, mobility becomes a crucial determinant because during the period of physical, imaginative and communicative immobility, migrants tended to imagine the homeland based on the experiences from the past, e.g. their subjective life-worldly perspectives on the locality and the Tamils living therein. These constructions of Jaffna remained important, because they structure the entire sequence of images which travelled communicatively and imaginatively from northern Sri Lanka to diaspora members thereafter. The relevant information was spread through official reporting on the war by foreign and local journalists, through the propaganda promoted by the LTTE, through the narratives of newly arriving diaspora members, through the information transmitted by those locals who managed to travel to the capital of Colombo and who had access to communication there, and through the few letters which could be exchanged infrequently. The sequential character of war history experienced from the viewpoint of the diaspora was thus shaped by the life-worldly images of co-presence in Jaffna during childhood. The images constructed from afar led to the emergence of so-called imagined spaces (Schiffauer 2006) in which memories and imaginations of the locality were conserved, transformed, and created anew. Although people followed the tide of the war and realized that changes had occurred, this kind of knowledge was not related to experiencing, or facing the locality and the people living therein. The new spaces filled with images from the past created by diaspora Tamils over time were detached from the proximate experience of the locality. The significance of the emerging time lag characterizing the Tamil diaspora's position towards the locality of origin for structuring social reality shows that temporality is an ordering principle of mobility, because it shapes the subjective and collectively shared images of human encounters, places, situations and moments which can be transmitted either through physical travel or communicative and virtual mobility. These forms of experience embedded in proximate and distant relationships thus are situated differently in time, but also in space.

Spatiality

Spatiality structuring the constellations of belonging and positioning of local Tamils vis-à-vis diaspora Tamils and vice versa is central, because spatiality is a determinant for the construction of transnational space. Migration and travel are specific forms of mobility (other than social mobility) which are always related to transcending geographical space. The migratory context of diaspora Tamils therefore entails spatial distance from the place of origin. In this regard, the case reveals how the significance of geographical space shapes belonging. Whereas Yuval-Davis Kanabiran and Vieten (2006: 7) focus on the argument that states and societies are affected by globalization in different ways with regard to belonging, I highlight the significance of geographical space and distance for belonging under conditions of mobility.

Travelling to Jaffna I was confronted with the desire to experience the territorial space, e.g. the place, in different ways. The Sri Lankan Army had declared some parts of Jaffna as so-called High Security Zones (HSZ) which were not accessible to civilians. The inhabitants of the villages located in these HSZs were displaced to camps, to relatives' houses, or they migrated. Even though the Ceasefire Agreement promised the relative stability of the peace process, the Sri Lankan Army refused to repatriate people within the army-occupied HSZs. Concerning this controversial issue one interviewee who originated from a village located in a High Security Zone described his claims towards the government: 'Before I die, can you please allow me to see my village?' This utterance makes it clear how important the emotional attachment to a place can be. In a similar vein Cheran (2007: 151) argues that in the Tamil tradition, exile or banishment from one's place/country/Ur[8] has always been a form of supreme punishment more severe than the death penalty. Furthermore, he relates this assumption to the diasporic space by claiming this to be a space of 'social death'. Cheran also argues that these patterns of deadly relationship to the place of origin are changing. He relates diaspora to embracing new possibilities for multiple forms of belonging and subjectivities, also signifying diaspora as spaces. Highlighting the ambivalent relationship between Tamils and their diasporic space, Cheran hints at the diverse ways of belonging as related to space and place.

On a more general level, Pfaff-Czarnecka (2008: 316) argues that distances and boundaries are not sufficiently taken into account in transnationalization research whereas spatial distances and political boundaries continue to affect migrants' existence in crucial ways. She traces this to the oversimplifying terminology of transnationality which deals with international problems in kinship and friendship relations. This is indeed the point I wish to make here: diaspora Tamils lived in a different location for many years, spending their everyday life and bearing the difficulties imposed by the receiving society. The forceful mundane dimension of living became localized in the recipient country. This process of localization in the receiving context changes the relationship to the place of origin, although diaspora politics, emotional bonds to the (imagined) homeland and the place of origin, including the people living therein, prevail as a salient feature structuring their life-worldly experience and contribute to the formation of transnational space. Nevertheless, limited opportunities for mobility and travel reduce transnational space maintenance to constructions based on communication and imagination. Consequently, the place itself is constructed only by means of imagination. However, the view from afar channelled through means of virtual co-presence is always selective and abandons worldly experiences embedded in geographical space within actual reach. In line with Cheran, I therefore argue that space and place are important for people's belonging and that the emotional attachment to a certain place changes over time. However, whereas he traces the changing modes of belonging to diasporic space, his argument is supported by stating that

---

8   The Tamil notion of Ur relates to place in the sense of village.

Tamil diaspora members' co-present encounters with their place of origin after the Ceasefire of 2002 dramatically changed the images of this place exactly because diaspora Tamils had appropriated diasporic space. This life-worldly appropriation of diasporic space while continuously constructing, imagining, maintaining and mobilizing for the homeland characterizes this particular Tamil translocal social space determined by immobility. Facing the homeland thus overthrow the old, yet salient, constructions of the place of origin and resulted in the necessity to redefine and reposition subjective and collective belonging within space.

Conclusions: globalization, (im)mobility and belonging

In this article I have discussed the relationship between mobility and belonging with regard to situations of proximity after a long period of immobility and absence. Analysing encounters of re-establishing co-presence between local and diaspora Tamils after the war-related isolation of northern Sri Lanka, I observed that despite transnational solidarity and the construction of global Tamil identity, differences between 'us' and 'them' became important. These constructions of difference only occurred due to the new opportunities for mobility after the isolation. The various dynamics of constructing sameness and difference are related to spatial and temporal dimensions and intersectionality and can be captured with the notion of belonging. A choice of identifications and identities situated in time and space are relevant if we want to understand the complex features of contact between the members of a transnational community or ethnoscape, who remained apart for a longer period of time.

Locating these findings and interpretations within the realm of approaches to (im)mobility, the significance of physical co-presence established through corporeal travel, or immobility, becomes particularly pertinent. In contrast to other forms of travel, corporeal travel enables the mobile person to have encounters which are qualified by the sensual experience of proximity. According to Urry (2002: 262), face-to-face contact, facing the place, and facing the moment is what characterizes proximate encounters. Proximity conceptualized in this way is interlinked with immobility (in the sense of mooring), because proximity presupposes dwelling in a particular place, at a particular moment in time, being involved in a particular face-to-face situation. Setting this observation into the translocal context stretched in time and space, proximity can only be achieved by corporeal travel. In the globalized world shaped by transnational migration and translocality, thus, corporeal travel turns into an experience entailing a particular quality which can not be achieved by virtual or communicative travel. The varying degree characterizing the intensity of experiencing other human beings, places and moments determines how human encounters resulting from mobility are shaped by positionality and belonging.

References

Adamson, F.B. 2002. Mobilizing for the transformation of home. Politicized identities and transnational practices, in *New Approaches to Migration? Transnational Communities and the Transformation of Home*, edited by N. Al-Ali and K. Koser. London: Routledge, 154–68.

Adey, P. 2006. If mobility is everything then it is nothing: towards a relational politics of (im)mobilities. *Mobilities*, 1(1), 75–94.

Anthias, F. 2002. Where do I belong? Narrating collective identity and translocational positionality. *Ethnicities*, 2(4), 491–514.

Appadurai, A. 1996. *Modernity at Large. Cultural Dimensions of Globalization*. Minneapolis, London: Minnesota University Press.

Basch, L., Glick Schiller, N. and Szanton Blanc, C. 1994. *Nations Unbound. Transnational Projects, Postcolonial Predicaments, and Deterritorialized Nation-States*. Langhorne PA: Gordon & Breach.

Bauman, Z. 2000. *Liquid Modernity*. Cambridge: Polity Press.

Brubaker, R. 2005. The 'diaspora' diaspora. *Ethnic and Racial Studies*, 28(1), 1–19.

Bruner, E.M. 1996. Tourism in Ghana: the representation of slavery and the return of the black diaspora. *American Anthropologist*, 98(2), 290–304.

Castells, M. 2000. *The Rise of the Network Society*. Oxford: Blackwell Publishing.

Cheran, R. 2001. *The Sixth Genre: Memory, History and the Tamil Diaspora Imagination*. Colombo: Marga Institute.

Cheran, R. 2004. Diaspora circulation and transnationalism as agents for change in the post conflict zones of Sri Lanka, in *Wenn es in der Heimat um Krieg und Frieden geht. Die Rolle der Diaspora in Krisenentwicklung und ziviler Konfliktbearbeitung*, edited by J. Calließ. Rehberg-Loccum: Evangelische Akademie Loccum, 263–80.

Cheran, R. 2007a. Transnationalism, development and social capital: Tamil community networks in Canada, in *Organising the Transnational*, edited by L. Goldring. Vancouver: University of British Columbia Press, 277–307.

Cheran, R. 2007b. Citizens of many worlds: theorizing Tamil DiasporiCity, in *History and Imagination. Tamil Culture in the Global Context*, edited by R. Cheran, D. Ambalavanar and C. Kanaganayakam. Toronto: TSAR Publications, 150–68.

Dannecker, P. 2005. Transnational migration and the transformation of gender relations: the case of Bangladeshi labour migrants. *Current Sociology*, 53(4), 655–74.

Elwert, G. 2002. Switching identity discourses: primordial emotions and the social construction of we-groups, in *Imagined Differences. Hatred and the Construction of Identity*, edited by G. Schlee. Hamburg: Lit, 33–54.

Faist, T. 2003. The border-crossing expansion of social space: concepts, questions and topics, in *Transnational Social Spaces. Agents, Networks and Institutions*, edited by T. Faist and E. Özveren Hauts. Aldershot: Ashgate, 1–34.

Fuglerud, O. 2001. Time and space in the Sri Lanka – Tamil Diaspora. *Nations and Nationalism*, 7(2), 195–213.

Gerharz, E. 2007. *Translocal Negotiations of Reconstruction and Development in Jaffna, Sri Lanka*. Bielefeld: unpublished PhD Thesis.

Gerharz, E. 2008. Opening to the world. Translocal post-war reconstruction in Northern Sri Lanka, in *The Making of World Society. Perspectives from Transnational Research*, edited by R.G. Anghel, E. Gerharz, G. Rescher and M. Salzbrunn, Bielefeld: transcript, 173–94.

Gunaratna, R. 1998. Impact of the mobilized Tamil diaspora on the protracted conflict in Sri Lanka, in *Negotiating Peace in Sri Lanka. Efforts, Failures and Lessons*, edited by K. Rupesinghe, London: International Alert, 301–28.

Hannam, K., Sheller M. and Urry J. 2006. Editorial: mobilities, immobilities and moorings. *Mobilities*, 1(1), 1–22.

Hannerz, U. 1987. The world in creolisation. *Africa*, 57, 546–59.

Hannerz, U. 1996. *Transnational Connections. Culture, People, Places*. London, New York: Routledge.

Harvey, D. 1989. *The Condition of Postmodernity*. Oxford: Blackwell Publishing.

Lachenmann, G. 2008. Researching translocal gendered spaces: methodological challenges, in *Negotiating Development in Muslim Societies. Gendered Spaces and Translocal Connections*, edited by G. Lachenmann and P. Dannecker. Lanham: Lexington Books, 13–36.

Maunaguru, S. 1995. Gendering Tamil nationalism: the construction of 'woman' in projects of protest and control, in *Unmaking the Nation. The Politics of Identity and History in Modern Sri Lanka*, edited by P. Jeganathan and Q. Ismail. Colombo: Social Scientists Association, 158–76.

McDowell, C. 1996. *A Tamil Asylum Diaspora. Sri Lankan Migration, Settlement and Politics in Switzerland*. Providence, Oxford: Berghahn Books.

Meyer, B. and Geschiere, P. 1999. Globalization and identity: dialectics of flow and closure, in *Globalization and Identity: Dialectics of Flow and Closure*, edited by B. Meyer and P. Geschiere. Oxford: Blackwell Publishing, 1–16.

Nederveen Pieterse, J. 1995. Globalization as hybridization, in *Global Modernities*, edited by M. Featherstone, S. Lash and R. Robertson. London: Sage Publications, 45–68.

Pfaff-Czarnecka, J. 2005. Das Lokale als Ressource im entgrenzten Wettbewerb. Das Verhandeln kollektiver Repäsentationen in Nepal-Himalaya. *Zeitschrift für Soziologie, Sonderheft 'Weltgesellschaft'*, 479–99.

Pfaff-Czarnecka, J. 2008. Are we all transnationalists now?, in *The Making of World Society. Perspectives from Transnational Research*, edited by R.G. Anghel, E. Gerharz, G. Rescher and M. Salzbrunn. Bielefeld: transcript, 311–24.

Pries, L. 1999. New migration in transnational spaces, in *Migration and Transnational Social Spaces*, edited by L. Pries. Aldershot: Ashgate, 1–35.

Rajasingham-Senanayake, D. 2004. Between reality and representation. Women's agency in war and post-conflict Sri Lanka. *Cultural Dynamics*, 16(2/3), 141–68.

Robertson, R. 1995. Glocalization: time-space and homegeneity-heterogeneity, in *Global Modernities*, edited by M. Featherstone, S. Lash and R. Robertson. London: Sage Publications, 25–44.

Scheyvens, R. 2007. Poor cousins no more: valuing the development potential of domestic and diaspora tourism. *Progress in Development Studies*, 7(4), 307–25.

Schiffauer, W. 2006. Transnationale Solidaritätsgruppen, Imaginäre Räume, Irreale Konditionalsätze, in *Die Macht des Lokalen in einer Welt ohne Grenzen*, edited by H. Berking. Frankfurt: Campus, 164–80.

Schutz, A. and Luckmann, T. 1979. *Strukturen der Lebenswelt*. Frankfurt am Main: Suhrkamp.

Spiegel, A. 2005. *Alltagswelten in translokalen Räumen. Bolivianische Migrantinnen in Buenos Aires*. Frankfurt am Main and London: IKO- Verlag für Interkulturelle Kommunikation.

Suryakumaran, C. 2002. *Kilali Crossing. A Tale of Despair and Desire*. Colombo: Vijitha Yapa Publications.

Tambiah, Y. 2005. Turncoat bodies. Sexuality and sex work under militarization in Sri Lanka. *Gender & Society*, 19(2), 243–61.

Tsing, A. 2002. Conclusion: the global situation, in *The Anthropology of Globalization. A Reader*, edited by J.X. Inda and R. Rosaldo. Malden: Blackwell Publishing, 453–85.

Urry, J. 2002. Mobility and proximity. *Sociology*, 36(2), 255–74.

Urry, J. 2007. *Mobilities*. Cambridge: Polity Press.

Wayland, S. 2004. Ethnonationalist networks and transnational opportunities: the Sri Lankan diaspora. *Review of International Studies*, 30, 405–26.

Yuval-Davis, N., Kannabiran, K. and Vieten, U.M. 2006. Introduction. Situating contemporary politics of belonging, in *The Situated Politics of Belonging*, edited by N. Yuval-Davis, K. Kannabiran and U.M. Vieten. London: Sage Publications, 1–14.

PART III
Global Firms/Urban Landscapes
as Scenery for
Proximity and Mobility

Chapter 6

Human Costs of Mobility: On Management in Multinational Companies

Laura Gherardi

Introduction to ambiguous mobility

One of the questions this book is concerned with is whether mobility today is a resource or a boundary. In this chapter, we will see that mobility can be both at the same time – specifically for the middle and top managers of multinationals who have international geographic mobility.[1] This is apparent from the results of the research we have carried out[2] on the patterns of international geographic mobility required in the work context regarding around 50 top managers and middle-managers, the majority of whom are from four multinationals, three based in Paris and one based in Milan.

Sociological studies of an empirical nature, specifically related to the mobility of social groups that possess important economic and cultural resources, are decidedly rare (see Sklair 2001, Wagner 2003). This is despite ubiquitous references to the growing mobility of people, goods and information, the opening up of

---

1 Namely that which, in Human Resources jargon, is called enterprise mobile population. Let us be clear that the groups on which we are concentrating, middle and top management, do not together include all those who are effectively mobile. For example, in the enterprise we have called 'multinational 1', there are technicians with a high level of specialization concerning one or more machines/production systems who transfer on a planetary scale for a period of time necessary for maintenance operations. There are experts or consultants who make transfers within international circuits and who, while not being part of employed staff, move within the enterprise for the period necessary for a meeting or a course. Moreover, there are salespeople and sector managers/area coordinators who move in different spatial scales, often internationally, or 'commuters' (researchers, for example, who make regular transfers, maybe for a week, between two or three laboratories of the same multinational situated in different countries of the same continent).

2 The research is part of a larger project analysing the relationship between mobility and power in the Italian, French and English upper classes as outlined in Gherardi 2009a. Part of the research carried out in the French territory was performed by the author together with Philippe Pierre, consultant and researcher at LISE/CNRS in Paris, previously Dhr l'Oreal Paris.

geographical borders and internationalization, as well as to the capacity of modern-day megalopolises to attract a population of business travellers. Sociological analyses on the geographical mobility of people, which deal with migrations on a global scale and aspects of exclusion and marginality on a metropolitan scale, focus on people who transfer for necessity – the poorly educated migrants for whom frontiers tend to close everywhere – or who remain trapped in their local area because they have no access to geographical or virtual mobility. Such analyses certainly carry out the important task of social criticism, introducing the idea that the geographic mobility of people, in the global age, is an extremely important factor of stratification which marks a first dividing line in the social space – which has on one side the mobile elites and on the other side those trapped in their local area. Castells (1996: 446) sums this up: 'elites are cosmopolitan, people are local'. Among sociologists of disparity, perhaps Bauman has provided the most explicit expression of this concept:[3] according to this author, the criterion that defines those at a high level and those at a low level in contemporary society is their degree of mobility; that is, the freedom to choose where to settle.

But the absence of empirical data regarding the mobility of those to whom sociology refers, in the fleetingness of a name – as *new international elites, the hyper-mobile, cosmopolites, the great globals, the cybernetic elite,* or *the global nomads* – without there being any agreement on the nature of the social groups involved,[4] has allowed the rhetoric of much of the most recent international management literature to monopolise the creation of the general perception of the subject. This is particularly true with regards to the international geographic mobility required of people in the working context. We use the term 'rhetoric' because, given the fact that the mobility of those involved is a strategic factor in supporting the spatial restructuring of capitalism in its current phase, current management literature concerning the organization of international mobility and

---

3   'Globalisation divides as much as it unites; it divides while uniting and the causes of division are the same which, on the other hand, promote the uniformity of the globe. In parallel with the emerging process of a planetary scale for the economy, finance, trade and information, there's also another process that imposes spatial barriers, what we call "localisation" (…) What appears as a conquest of globalisation for some, represents a reduction in the local dimension for others; whereas for some globalisation means new forms of freedom, for many others it feels like an unwanted and cruel destiny. Mobility has moved up the ladder of values that create prestige, and freedom of movement, always a rare commodity and distributed in an unequal manner, is rapidly becoming the main factor for social stratification in our times, whether we call it post-modern or late modern' (Bauman 1999: 4).

4   This is a significant and open issue in contemporary sociology. See, for example, Webster 2002.

of intercultural management[5] treats geographical mobility as a value in itself.[6] After having been thematized for two centuries, from literature of the romantic period to the 1970s, first as an oppositional factor to the spatio-temporal discipline imposed by the bourgeois industrial order in its early days, and then by production in its Fordian phase,[7] mobility appears in recent management literature as the main characteristic of the powerful and as a mainstay of the global image they propagate. The architecture of such imagery – emblematic of the New Spirit of Capitalism[8] – rests on the commendation of a totally positive geographic mobility, at no personal cost to those who live it. Such a mobility which is said to link together a homogeneous élite of citizens of the world, of explorers of culture for whom mobility is always associated with personal realization.

Although this imagery can be perceived as common sense, we will see in the following paragraph that a very different picture, point by point, arises from the findings of an analysis of the practices of mobility and of the relative attitudes of management involved in intense international mobility. Alongside those whose only regret is not to have started travelling earlier, and who emphasize the enriching aspect, the continuous excitement of mobility, we have met people who talk of their mobility as an obligation, a test, a sacrifice. More often, in discussions with such interviewees, mobility varies between a privilege and a price to pay, a vehicle of power and an imperative which carries heavy personal costs. This is ambiguous mobility, which, in practice, is a functional mobility, with tight rhythms, dictated by the needs of production, unlike romantic mobility – promoted by management literature and the aim of which is personal realization. The functional mobility required of our interviewees, particularly top managers, fighting against time, seems instead to impede their self realization in certain aspects of their lives. 'Every week for twenty years I travelled the infernal circuit between three international locations, then I gave it up and within two months my daughter, who is thirteen, said to me, "I discovered I had a father"' (Ex-director general of a French multinational).

---

5   The field of intercultural management deals principally with themes such as negotiating beyond cultural barriers, the management of multicultural teams, training future expatriates and the management of mergers or of international group companies.

6   In this article, the reference is to a review of the most recent management literature, composed of around 40 texts in Italian, English and French, published from 2000 onwards. We would, however, like to point out that Boltanski and Chiapello have found that mobility was already lauded in managerial theory in the 1990s. According to these two sociologists, the ideology supporting capitalism has incorporated some of the values – such as authenticity and autonomy that are linked to mobility – in the name of which capitalism itself was criticised in 1968 (Boltanski and Chiapello 1999).

7   See previous note. On the adoption of mobility in the axiology of advanced capitalism through the artistic criticism of May 1968, which in turn had its roots in the Bohemian Paris of the nineteenth century, see Boltanski and Chiapello 1999.

8   That is the ideology that justifies and motivates commitment to capitalism, see Boltanski and Chiapello 1999.

The personal costs exacted by the mobility that has become the norm in advanced capitalism from the lives of the middle-managers and top managers interviewed appear to derive mainly from the rhythm of their movements. It is this rhythm which characterizes the different recurrent and distinct forms of mobility that we have identified and which, taking into account the power differentials, militate against the myth of a homogeneous élite of citizens of the world. In particular, we will focus on the social threshold evidenced by the difference between the international circulation of top management and the mobility of the economic élites of long standing which financial capitalism has liberated from all spatial constraints and which, therefore, produces the only social group able to choose freely between mobility and immobility.[9] Furthermore the rhythm of movements is significant in that, as we shall see, it dictates the conditions of the relationship that mobile individuals can establish with the places they visit and with the cultures they come into contact with.

Forms of mobility in spatial flow

From the first interviews and from the analyses of mobility planning within our sample, differences became apparent, not only in the direction of the movements made by the subjects, but also in the rhythms of the movements themselves in terms of duration and frequency. By combining time and space, this rhythm allows the individuals' mobility not to be reduced to a series of lines drawn on the map of the world, unlike management literature, which detaches space from time, the subject from the organization he works for and the position he occupies, and his movements from the reason and logic inherent in them. Rhythm allows to consider individuals' mobility in terms of spatio-temporal configurations that are recurrent and distinct, and that constitute the many forms of mobility in our sample. Specifically, the next subparagraphs will deal, respectively, with the expatriation of middle management; with the form of mobility we have called reticular mobility with single base; and with the international circulation of top management which we will compare with the chosen mobility and cosmopolitanism of one section of the economic élite of long standing.[10]

Space-time configuration 1: expatriation

The first space-temporal configuration described characterizes the form of mobility referred to as expatriation: this involves a transfer from head office to a foreign

---

9   The interviews we give extracts from in the section 'Space-time configuration 3' were conducted between the writer and representatives of the economic elites in the context of the wider research referred to in note 2, in which those inheriting large fortunes in land and moveable property were included in the larger sample.

10   See previous note.

branch of the company, usually for a period ranging from two to five years. Thanks to its specific configuration, this is the only form of mobility that we are analysing for which it is possible to find a minimum amount of bibliographical reference[11] and it could occur for numerous reasons, following, for example, mergers or acquisitions of a foreign enterprise or the opening of a branch in a foreign land.

The long period of stay in the place of destination obliges the expatriated middle-managers to put down new roots and to live their daily lives in a different culture from the country of origin, adapting to the new codes linking the places, people and objects available, or dealing at a distance with people, places and 'anchors' in the country of origin that often take on a connotation of 'meaningful elsewhere' (Ramos 2006: 58). The result is that the people involved are 'here' in the place of destination and 'elsewhere', in the country of origin, (and the reverse happens when they return) or, more often, they are constantly between the two. References to places and facilities are often duplicated in the discourses of the people belonging to this group: 'my favourite restaurant here in Paris is X, in Turkey it's Y, a little pink house on the Bosphorus, it's not a chic restaurant, but it's mine, we often go there, it's my fishermen's restaurant on the Bosphorus' (middle-manager expatriated from Paris to Istanbul for four years). As a matter of fact, the attempt to put down roots – which when achieved can, in fact, lead to a permanent transfer to the destination country, or to a sequence of expatriations, one after the other, to different countries – appears more often than not to begin with the identification of 'corners of the city' and 'facilities' that become the first points of reference for the creation of a new routine, of a new *chez soi*.

These people, who for the most part are men,[12] are individuals possessing significant economic and cultural resources. They approach the problem of the services offered by the new city in which they live for a long time and the multiplicity of alliances and the creation of an identity in different cultural contexts in a new way compared to disadvantaged immigrant workers. The costs which the cultural differences may inflict on individuals in terms of identity when expatriation turns out to be an unhappy experience, can create the feeling of a prolonged period of sufferance. In management literature, difficulties, when mentioned, are described, at the most, as a break from normal daily routines (Black and Gregersen 1999), as effects of cultural shock, and there is no reference to conflicts of identity experienced by people in the name of a business culture which is supposed to remove these conflicts on its own. In other words, the enterprise's illusory undifferentiated citizenship would here cancel any need for 'putting down roots', particularly ethnic national roots, while our observations reveal that the different national identities of individuals that come into contact with one another in multinational companies produce non-homogeneous practices and behaviour within middle-management, which is a favoured target of intercultural management. The international does not abolish the national, in the sense that within transnational space, nationality

---

11   See Cerdin 2007.
12   On the male domination in this field, see Gherardi 2009.

of origin is an element that 'stratifies' the population, such that the international management space remains heavily hierarchized by the importance of global business exchange.[13]

With mobility offering economic incentives and promising career advancement on return, it assumes the nature of a test. With mobility being considered a sign of such qualities as openness, curiosity, love of diversity, self confidence, relational capacity, enthusiasm and self denial for the sake of one's work,[14] a refusal to move, whether it be to leave in the first place or to undergo further moves abroad once having returned, is deemed an expression of the opposite values. Thus mobility strays onto the ethical plane, it becomes a social precept, whereas for real freedom of choice, the options of mobility or immobility should have equal value. An unsuccessful move abroad lead to marginalization of the employee within the company, while a successful move abroad is significantly appreciated, within the labour market, only when its destination is an industrialized country. But mobility is also a test because it impinges on the family life of the employee: the choice may be one of transferring the whole nuclear family to the country of destination – meaning that it could impact upon the career of the spouse, even where the company of the expatriate (who is generally male) undertakes to provide local work for her, and possibly on the careers of any children[15] – or one of waiting, or even one of breaking up: 'My family life is mobile too! I got divorced because my wife didn't want to move, she wanted her career to be in France and so didn't come with me to Ireland, she is static and has always lived inside the same three square kilometre area' (Expatriate manager, multinational, based in Paris). Adaptation seems, therefore, to depend largely on the ability to change one's point of view and way of thinking while maintaining one's feeling of self and a sense of continuity in one's family situation.

---

13   As shown by Wagner (1998), in transnational spaces, nationality of origin is an element that 'stratifies' the population.

14   The distinction between 'stay-at-homes' and 'expatriables' is a moral one: unlike closing up, obtuseness, the xenophobia attributed to 'locals', 'sedentary' or 'rigid' people (in the sense of being attached to ideas or values that are not liable to change), mobility is considered in itself to be proof of an openness of spirit, tolerance, the ability to adapt and the valuing of difference. International managers whose experiences have been successful sometimes take this view: 'I am different from the people who have not moved away, my mind is open and theirs are closed' (expatriate from a large Italian company, three years, Milan-Barcelona).

15   For example, Global Nomad International USA, formed in 1986, is concerned with the particular problems that can be experienced by children and adolescents who have grown up in more than one country due to the long term professional mobility of their parents; problems related to bonding – from 'affective inclination' to a real emotional anaesthesia, an inability to love – and to the dynamics of attachment-abandonment-substitution in their social relations.

Space-time configuration 2: reticular mobility with single base

The second form of mobility identified is what we have called reticular mobility with a single base, since it involves frequent transfers from the base, usually on a monthly or two-monthly basis, generally for a period ranging from a few days to a few weeks, generally from the country of origin towards different locations abroad, plus the return journey. This spatial-time configuration characterizes, for example, the mobility of international project managers and appears to be the most variable configuration since the number and direction of transfers change, also with regards to the same individual, depending on the international projects launched by the multinational firm. We give the name of *international managers* to middle-managers who, having this sort of mobility, essentially have the task of controlling and coordinating multinational teams that are not hierarchically linked to them. The multiplicity of the contexts that the international middle-managers have to physically deal with means that they often find themselves 'caught in the middle', which may pose a problem of adaptation and multiple identity. While repeated mobility is considered by the enterprise as a process that allows the individual to gradually improve their ability to adapt, when the rhythms are particularly fast and 'over-circulation' of the individuals in geographic space occurs, the motivation associated with transfers appears to collapse for our interviewees. The short duration of the stays in their destination countries does not, in general, enable them to get to know the locality beyond the better known tourist locations – 'I was in Japan for five weeks, it's not like Vietnam, I was in Vietnam for three days, in Japan I walked the streets, I went into shops, I had some knives made, I went to the Temple, I passed the time pleasantly in the main square' (Director of business education and development of worldwide operations for a multinational company based in Paris). In conversations with the interviewees in this group, mobility often swings between a choice that reflects their attitudes, a resource, in the sense of a career accelerator, and an imperative, in the sense of a *necessary investment* they need to make. These words from a young international manager from Milan are an eloquent example: having confirmed himself 'to be made for travelling, for exploring', and hence 'at ease anywhere', he told us that 'travelling is stressful' and that since undertaking international travel he has put on so much weight that he is unrecognizable, because 'when you live away from home you need to comfort yourself, you have to spoil yourself and get your body accustomed to resisting change.'

The promotion and management of mobility among such employees has assumed great importance commensurately with the management of the knowledge that is built up through the interrelation between decision support systems, HR management and formal communication channels. For this, companies promote mobility as an alternative to normal career paths and seek to provide good examples:

> The company wants you to keep on the move, and if you want to get on in your career it's often the only thing to do. Let me give you an example: we have some product marketing managers who all start at the same time with the same qualifications, the same language abilities and the same areas of study, one of them becomes head of group and one goes to work abroad. Which one is going to be marketing director in three years' time? The one who went abroad. Why? The company's answer is that he has better knowledge, vision and strategy and he understands company decisions, he has learned the 'core business' of the company and is valued more than the person who stayed put, in people's minds he became director because he went abroad (…) But, unlike us, the company doesn't understand the sacrifices involved in travelling. (Expatriate manager of a large company based in Paris)

While some people tell us they experience a sense of having time out, for example while taking solitary evening walks, for others, or for the same people at different times of their lives, a sense of insecurity prevails: 'When you travel, you're completely independent, you're completely alone, you're not completely safe because if something happens to you, you can't call anyone. You have to be strong, and there are times when you don't feel like being fully alert and you can relax' (Project manager of a multinational company based in Milan).

In some cases, the tension between freedom and dependency, between adaptation and permanent self-identity can threaten the ontological state: 'I sometimes feel anxious. When I'm half-asleep I wonder where I am and I have to repeat to myself, for example: you're in Milan, you're in Milan, the city will come to life soon. You're in Milan, you're in Milan, and the city will come to life soon …' (Project leader international operations, large Italian company). In some cases, we are really a long way from the figure of the 'transnational manager' described in management literature as a person perfectly at ease with the ordeal of repeated journeys, for whom uncertainty is the norm and who, at this point, represents the ideal light and mobile global player since, according to management theory, he doesn't have any roots: 'My companion is also very mobile, but I hope one day to be able to start a family, and in that case we will both stop; we can't go on like this if we want a family'. (Project manager of a multinational company, based in Paris).

International resources, passed on from the family of origin as well as scholastic institutions, have been found to be very important for this form of mobility. The resources to which this group of interviewees make constant reference include, for example, knowledge of languages, the habit of frequent travel, the opportunity to develop relations with individuals of different nationalities and a *savoir faire* in different contexts. The following extract – taken from an interview we had with a French international middle-manager, son of an army officer who came into contact with profoundly different, and at times contradictory, contexts of socialization – is an eloquent example, albeit a limited case:

I have [frequent travel] in my blood (…). I was born in Algeria. From the age of five to ten I lived in Colorado without ever returning to France. I spoke English better than French. Then I moved to Chambéry at the end of 1968 and I was considered in the Alps as an American, a dirty capitalist, the son of a soldier in a working-class environment. After that I went to London for three years where I was invited to punk gatherings, while in Chambéry I went to housewives' afternoon teas at 4 o'clock in the afternoon. This gives you an international outlook because you're forced to adapt. I'm used to change because I was trained as a child. If you teach your child to eat raw worms with a bit of vinegar when he's one year old, he'll think it normal when he's thirty. (International middle-manager of a multinational company based in Paris)

Space-time configuration 3: circulation of externally mobile top managers. Internationalism of top management vs cosmopolitanism of the economic élite

The third spatial-time configuration, which we have called circulation of externally mobile top management, is characterized by transfers which are very brief, of one or two days, and regular – from one to three international transfers per week. This pattern seeks to reconcile two opposite requirements of the top managers who we met. That is, the requirement not to leave the head office for too long and the need to build or consolidate partnerships with external organizations or individuals (whether these are customers, suppliers, shareholders, etc.).

Among the costs described to us there is the fatigue of sustaining the rhythms of the transfers, physical or psychic fatigue, and more often both:

It's a life of permanent transfers and therefore also of permanent time changes. You're constantly on the go. You're never able to rest completely. The result: we're permanently tired (…). There are some who, it's true, take tranquillizers and stimulants to keep them going, or maybe a few drinks. (Sales director of a multinational company based in Milan)

There is never, however, any mention of adaptation to different cultures from the country of origin, nor any identity problems – 'I certainly feel French. Should I feel English because I speak English for my job?' (General Manager of a multinational company, based in Paris) – certainly due to the brevity of the time spent in the country of destination (see Kaufmann 2005). This is also because the fast rhythms impose a choice of locations which are purely functional, such as hotels near the airport, and which do not allow for any form of putting down of roots, nor the possibility of visiting the sites where these people stay, as this former general manager of an Italian multinational explains:

… I organised a road show, a method which, from a business community point of view, means meeting financial analysts from the countries which have shareholdings in the holding company. Specifically, I went to the City in

London (…). I only saw the City during transfers with the chauffeur, from the car window. I saw a bit more in Paris because they drove me through the historic part of the Défense, so I saw a bit more.

Or as the General Manager of a large French company explains:

I do one journey per week if it's long distance, or two if it's in Europe – Germany, Switzerland, Italy or England – always short two-day return journeys, while at the same time trying to arrange as many meetings as possible, which means leaving very early in the morning and returning very late in the evening so you can do as much work as possible during the day, and maybe spending a night or two at the workplace: when you travel in these conditions, you don't have time to see things, I've been to Frankfurt maybe fifty or sixty times and I've never been to the old town to see Goethe's house.

What is more, there is a very tight organization of the time during the transfers[16] and the need, for most of the top managers we interviewed, to be constantly 'connected': 'When you arrive somewhere, you get off the plane, you connect the telephone, the blackberry, and you receive e-mails. We're therefore always connected and obviously the flow of e-mails, which is a constant stream, increasing the stress even further, because we see the mail arriving and we have to write replies and so on' (General Manager of a multinational company based in Paris).

What is described above explains in part why practically all the people in this group expressed a disenchanted attitude towards the link between mobility and personal achievement, and between mobility and freedom from control. Again in this case, the costs take a toll on family life:

I have a big family, but time is a tyrant. This is possibly the worst aspect. I've never even known which school my daughters attend. When I bought a beautiful villa, I was in the United States. They organised the move without me. I arrived and the house was ready. They still harp on about it. What they say is true: you live an unreal life. (Sector manager of a multinational company based in Milan)

It is important to point out that this last form of mobility, which is strictly functional and with very fast rhythms, is very different from the mobility of the economic élite in terms of location, rhythm, resources and motives. There are no hotels, no

---

16   To give another example: 'I give you an example, we arrive in USA on Wednesday at 7pm, we are tired, we try to perform until 10pm, the day after at 8am we start our meetings all day long, we finish at 6pm, we go to our hotel to start working again, some mails need an answer. Again, we work the all Friday and as soon as we finish we take the plane in order to be in Paris on Saturday morning' (Communications Director of a multinational based in Paris).

company offices but instead, clubs, private villas or castles and diplomatic offices are the places in which this group of the economic elite usually moves. The weekly transfers of Anthony O'Reilly, published in *Newsweek*[17] are a good example; he flies from Dublin to London for a work lunch and to Rome for a dinner at the English Embassy, and then goes to Normandy at the Château des Ducs in Dauville, one of his homes, which once belonged to William the Conqueror. One of the investors we interviewed explained to us the difference between the migrations of top management and his nomadism based on the self-determination of rhythm and destinations of his transfers. This brings to light a further aspect of the effective multi-territoriality shared by many of the people belonging to this social group:

> I have a house in London, a telephone in London, an Internet connection in London, I pay my bills in London, I have a car in London, and then I come to Milan. I have a house in Milan; I have a car in Milan, and so on. I have a bank in Milan, a bank in London, a bank in San Francisco, a bank in Australia; that is, wherever I am. I go out and I've got my bar where I have a coffee. I haven't got a house in Paris because I don't like Paris. I'm not interested in French people. You set up the homes and then do the business. Otherwise we're back at the beginning. First you get involved in something that you enjoy. Then you see what happens. You have to follow your irrational impulses as much as possible. I never stay for years in the same place. It's impossible. It's because I'm a nomad; it's not for work. I could take years of sabbaticals ... I organise the meetings. I can assure you that no one arranges my time. (Hi-tech investor, heir of a great fortune, Milan)

It is, therefore, a disinterested mobility which follows a different logic from the productive logic implied by the international circulation of management: 'Managers who do normal work like that of secretaries move like whales, in San Francisco whales pass by in March on their way north in search of colder seas, then again in September or October on their way south, while the nomad moves around, he doesn't necessarily go back the same way, he doesn't move between fixed points from A to B to C' (Hi-tech investor, heir of a great fortune, Milan). It is a mobility associated with a particular form of self realization,[18]

---

17   *Newsweek* 15/22, May 2006.

18   'The objective of my nomadic life is to construct the self, or better, to reveal my true self' – here we see a reprise of the dual concept of mobility-realization. 'There's a caption on an enormous billboard I always see when I arrive in San Francisco by sea which says: "You never risk losing yourself until you stop", which means: when you run you can't get lost, it's when you stop that you find out you're lost, the thing is to keep going; I have come to understand more and more that I am the nomadic type. The important thing is not whether I'm a giver or a social parasite, it's that I'm living this kind of life, I feel different, I'm living the Gauguin myth. He moved from Paris to Tahiti, that's not a bad idea' (Hi-tech investor and heir of a great fortune, Milan).

sustained by a multiterritoriality which enables élite's members to feel at home in different countries, and which therefore avoids the costs of dislocation. A comparison between the circulation of management and the mobility of a section of the economic élite, particularly one of long duration, reveals a social boundary, marked by the unequal availability of time and space between the two groups and, more fundamentally, by a different kind of freedom in the choice between mobility and immobility. None of the business heirs we came across working in large family-based companies spoke to us about geographic mobility in the working environment as a cost. Rather, it serves as evidence of considerable independence from the rhythms of production, as can be seen, for example, from the words of this interviewee, president of two companies belonging to the large family business group of which he is vice president: 'For me the journey is a time of absorption, I make my journeys by day, I like them, as long as it doesn't get too late, that destroys you and when you start early the next morning you are tired, the tiredness stays with you all day and you never quite recover'.[19] Furthermore, to understand the diversity of management's relationship with foreign countries and 'the aristocracy of money' (Pinçon and Pinçon-Charlot 2007) it must be mentioned that the exponents of the latter, as well as inheriting business capital, which is today entirely convertible into financial capital,[20] more often than not also inherit a cosmopolitan capital. That is a social capital of international relations with members of the élite of other countries – with whom contact is generally facilitated both through the ownership of homes or *pieds à terre* abroad and through membership of exclusive circles. This diversity has its roots in the family and in socialization from childhood: 'The cosmopolitanism of the old upper classes is a borderline case towards which international management culture leans without the two relationships becoming combined abroad' (Wagner 1998: 87). While internationalism is a fairly recent phenomenon, the cosmopolitanism of the economic élites, which is essentially supported by very tight but at the same time geographically scattered family and social networks, is by no means new.[21]

---

19   This person refers to a day which begins with 'two hours of leisure in the morning, completely alone doing some light reading, maybe a newspaper or some magazines I skimmed through the night before, or maybe some work papers' and usually ends 'in the late afternoon, when I go back home to my family'. The naturalization of privileges, which in this case means a high degree of freedom to manage your own time and space, arises from reducing differences in experience to a question of temperament and values, while denying the social mechanisms that enable them: 'Lots of managers have to stay away from home for long periods, although I should say I know a lot and they seem very relaxed and happy, I couldn't do it, unless I'm travelling during the day I do very short journeys, never more than one or two a year, because as I said I like to be with my family'.

20   Financial capital permits the highest degree of independence from the rhythm of production.

21   It can be seen, for example, in the interweaving of diplomatic relations between the dynastic families of the past, and in the growth of high finance through the operations of the big families described by Bergeron (1991).

Cosmopolitanism, the result of the accumulation among élites of extranational capital over several generations, shows itself in a milieu of which it is a strong identifying element, while international resources, in the composition of which the family also plays a part,[22] can be acquired by individuals while still in education and during their professional lives. This essentially by virtue of the geographic mobility practised by institutions and, in this way, resourcefulness is revealed through mobility.

Conclusions

Our analysis underlines the problematic nature of romantic mobility in advanced capitalism, that is to say of a totally positive image of mobility as seen in management literature as an end in itself, and on the possible relationship between the geographic mobility required in a working environment and personal fulfilment. We have shown how the international circulation of top management to very tight rhythms reveals the low level of freedom this socio-professional group has in organizing their own time and space and how such subordination can produce the violent repercussions, particularly on personal relationships, described by the interviewees. A dark side has thus emerged – unexpressed, because mobility is eulogized in management literature as a resource that requires no reward – of a mobility that is at the service of production. According to the logic of continuous performance, mobility often changes from being a stimulating factor to one which contributes heavily to the causes of depression[23] and burnout. This is also true of the middle-managers interviewed, whose forms of mobility – expatriation and reticular mobility with single base, which also produce gentler rhythms – can extend beyond the emotional sphere, and in particular upon the identity-making sphere in its strict sense. Put into the service of the very forces it was meant to destroy,[24] mobility is full of ambiguities. Besides being a factor of inequality at the lower end of the social scale, towards the high end of the professional hierarchy, for some socio-professional categories, it is a resource which becomes an imperative, a criterion for selection and testing – which does not provide freedom from control, but instead increases temporal restrictions. It sets up a new ethical norm, a requirement which, in a way that is unexpressed, produces tension between ties and risks, between confidence and emotional anaesthesia, between being uprooted

---

22   As we have also shown with the example of the international manager who is the son of a soldier.

23   In this respect, see also the reflections of Ehrenberg (Ehrenberg 1991, 1999, 2000) on depression as a contemporary social illness of time and space.

24   As indicated in the introduction, historically mobility appeared, in particular by romantic literature, as an element of personal realization and freedom from the capitalist universe, especially from stability and from the rules imposed by the bourgeois industrial order.

and being secure, between autonomy and dependence. It affects the balances that people try to set up between the various areas of their lives and within themselves – in particular, on the formation of ties and, perhaps in a deeper sense, on the development of their identity which is bound up with the space and time in which their lives are played out.

The ambiguous nature of the mobility required of our interviewees was particularly revealed in the comparison between the international circulation of management and the mobility of one section of the economic élite of long standing which, mostly thanks to the independence it enjoys in relation to the rhythms of production, can determine its own time and space, hence its own mobility or immobility – in each case maintaining the ability to dictate the rhythms of others (cf. Gherardi 2009b). In the case of the moneyed aristocracy, the threshold which separates them from management is based on a dual form of inheritance, that of business capital and cosmopolitan social capital. As if to say that space can bear traces of the social mechanism of the reproduction of privileges, and hence of accumulation between generations – this is another possible link between time and space in addition to the inseparable nature of the two dimensions in the rhythm of movement encapsulated in the expression 'spatio-temporal configuration'. The identification of different forms of mobility based on inherent differentials between those who travel across the spatial flow militates against the myth expressed in management literature of the existence of a mobile élite that is socially homogeneous.

References

Bauman, Z. 1998. *Globalization. The Human Consequences*. Cambridge, Oxford: Cambridge Polity Press/Blackwell.

Bergeron, L. 1991. *Les Rothschild et les autres*. Paris: Perrin.

Black, J.S and Gregersen, H.B. 1999. The right way to manage expats. *Harvard Business Review*, 77(2), 52–63.

Boltanski, L. and Chiapello, E. 1999. *Le Nouvel Esprit du Capitalisme*. Paris: Gallimard.

Castells, M. 1996. *The Rise of the Network Society. The Information Age: Economy, Society and Culture*, Vol. I. Oxford: Blackwell.

Cerdin, J.L. 2007. *S'expatrier en toute connaissance de cause*. Paris: Eyrolles.

Ehrenberg, A. 1991. *Le culte de la performance*. Paris: Calman-Lévy.

Ehrenberg, A. 1999. *L'individu incertain*. Paris: Hachette.

Ehrenberg, A. 2000. *La fatigue d'être soi. Dépression et société*. Paris: Odile Jacob.

Gherardi, L. 2009a. *Mobilité ambigue. Pour une sociologie des classes sociales supérieures dans la société contemporaine*. PhD Thesis. Paris: EHESS-UC.

Gherardi, L. 2009b. Ereditare la libertà sul proprio tempo e spazio e il potere su quello altrui. *Dialoghi internazionali*, 10. Milan: Mondadori.

Gupta, A.K. and Govindarajan, V. 2000. Knowledge flows within multinational corporations. *Strategic Management Journal*, 21, 473–96.

Harvey, D. 1989. *The Condition of Postmodernity*. Oxford: Oxford University Press.

Hirschman, A. 2007. *Exit, Voice and Loyalty*. Cambridge MA: Harvard University Press.

Kaufmann, V. 2005. Mobilités et réversibilités: vers des sociétés plus fluides? *Cahiers Internationaux de Sociologie*, 118(1), 119–35.

Pinçon, M. and Pinçon-Charlot, M. 2000. *Sociologie de la bourgeoisie*. Paris: La Découverte.

Pinçon, M. and Pinçon-Charlot, M. 2007. *Les ghettos du gotha*. Paris: Le Seuil.

Ramos, E. 2006. *L'invention des origines. Sociologie de l'ancrage identitaire*. Paris: Armand Colin.

Sennett, R. 1998. *The Corrosion of Character: Personal Consequences of Work in the New Capitalism*. New York: W.W. Norton and Company.

Sklair, L. 2001. *The Transnational Capitalist Class*. Oxford: Blackwell.

Tarrius, A. 2000. *Les nouveaux cosmopolitismes, mobilités, identités, territoires*. Paris: Editions de L'Aube.

Urry, J. 2000. *Sociology beyond Societies. Mobilities for the 21st Century*. London: Routledge.

Wagner, A.C. 1998. *Les nouvelles élites de la mondialisation. Une immigrée dorée en France*. Paris: PUF.

Wagner, A.C. 2003. La bourgeoisie face à la mondialisation. *Mouvements*, 26(2), 33–9.

Wagner, A.C. 2007. *Les classes sociales dans la mondialisation*. Paris: La Découverte.

Webster, F. 2002. *Theories of the Information Society*. London: Routledge.

Chapter 7

Urban Mobility, Accessibility and Social Equity: A Comparative Study in Four European Metropolitan Areas

Matteo Colleoni

Introduction

An increasing number of studies in the last two decades have dealt with the theme of the interactions between the spatial and morphological features of the city, socio-economic structure of the societies dwelling in them and mobility of populations. Many of these studies have concluded that mobility is growing more and more substantial in urban societies, with increasingly less marked divergences between genders, professions and income classes, and that suburban areas, and the populations living in them, are the most affected (Mogridge 1985, Newman and Kenworthy 1989, Naess et al. 1995, Fouchier 1998, Mo.Ve 2005).

The explanations relate to the changes taking place in the labour market, in the more general socio-economic system and in urban morphology, in particular to the progressive concentration of workplaces (above all administrative and management functions) and of opportunities (urban resources, services and assets) in the central areas of cities, and the location of residences in increasingly distant and scattered suburban areas (Martinotti 1999, Schwanen et al. 2001, Stead and Marshall 2001, Naess and Jensen 2004, Naess 2006). The resulting spatial and social spread of mobility has been described as the consequence of the gradual acquisition of the right to travel by increasingly numerous social groups, and in its turn to the condition for their improved accessibility to urban spaces and opportunities.

Although mobility is an essential condition to access urban assets and services and a fundamental factor for social and urban integration, it is however not evenly distributed between individuals and social groups and, above all, does not everywhere have the same quality relating to the resources used and the restrictions limiting their use. The inequalities relate both to the different social distribution of access resources (urban, socio-economic, cultural and temporal) and to the presence of restrictions (space-time and social) which hinder their use.

The objective of the study presented is to analyse the way in which:

- the urban structure of residential areas influences the presence and availability of opportunities;
- the location of residential areas and opportunities, combined with the different ownership of mobility capital by residents (Kaufmann et al. 2004),[1] influences mobility styles and accessibility to urban assets and services.

In detail, the study relates to a comparative European research carried out in four metropolitan areas by Mo.Ve[2] and its partners in 2005–6.[3]

Theoretic references

Access to urban assets and services, the so-called opportunities of which Anglo-Saxon urban sociology speaks (Dijst et al. 2002, Urry 2002), has always depended on the morphology of human settlements; of residential ones, firstly, but also of workplaces and services.

For many centuries we lived in cities with a compact, densely constructed morphology around historic urban centres, particularly in European countries with a high level of development, in which residences, work places and services were close and the identity of populations was based on belonging to the local communities of family relations or neighbourhood. In this settlement model, travelling and accessing services, was relatively simple, firstly for space-time reasons, since the local assets and services were close to homes, but also for cultural reasons, since the universal right of citizenship guaranteed the principle of access to public space and its resources (apart from the limits posed by inequalities of class, religion and local belonging).

This did not change even when the urban development process led to the birth of suburban districts in which, despite the presence of a lower number of assets and services, with less specificity and a more marked mono-functional nature, access was nevertheless made possible by their relative space-time proximity, by the presence of transport services and, above all, by the spread of the use of cars.

The situation changed with the birth of the so-called diffused (or limitless) city, in which the peri-urban[4] became the privileged location area for settlements

---

1   The term mobility capital has been used, among others, by Kaufmann to indicate the set of resources (called access resources) owned by an individual and facilitating their mobility and access to urban assets and services (Kaufmann et al. 2004).

2   Mo.Ve is the international Forum on sustainable mobility in European metropolitan areas. Its products are consultable on the Internet site www.move-forum.net.

3   Mo.Ve's partners include the Department of Sociology and Social Research in the University of Milan Bicocca.

4   By peri-urban we mean the area of settlements stretching between the boundaries of the historical cities and the low-density area, at times inappropriately called countryside (Martinotti 1999).

and radically changed the space-time morphology of mobility and accessibility to urban assets and services.

The reasons for the process of urban spread are various:

- the progressive loss of importance of the primary economic sector and the availability of former agricultural land for building at lower prices than in urban areas;
- the decision made by young families to move to centres with lower demographic densities, attracted both by lower housing prices and by higher quality residential contexts;
- the transfer first of businesses and then of large-scale retail distribution services to the outskirts of cities provided with large areas for development and, above all, better serviced by infrastructures for moving goods and persons.

The process of urban spread had different times and took on different forms in cities in Western countries. Unlike what happened in cities in the United States and in many other European cities, in Italy urban spread was the result, almost always unplanned and unexpected, of the strong development of medium-sized inhabited centres rather than the result of their agglomeration within the increasingly spreading boundaries of large cities and metropolises. The resulting scenario was the urban continuum which may be seen in the megalopolis in the Po Valley, but also along the urban artery joining the capital towns in the Veneto and Emilia regions, whose boundaries often do not match those of the pre-existing metropolitan areas, being shaped more like corridors, places for residential settlement, production and service and increasing flow spaces (Castells 1996, Martinotti 2004).[5] Within the new linear urban systems the scattering of settlements is not spread evenly, through a process of extension from the metropolitan centres, but along the lines linking the urban areas distinguished by the presence of complex functional interactions rather than simple relations of domination or dependency (Bagnasco and Le Galès 2001, Sassen 2001).

The spreading of settlements has brought with it the increased demand for mobility and caused the crisis in the traditional system of public transport supply which, organized on the premise of the compact city, has a mainly radical structure, lacking in extra-urban transport networks and centres for modal interchange. With due distinctions related to the type of national and regional context, this situation

---

5   The inadequacy of the terms in use to define the new urban systems has prompted some authors to adopt the term mega-city, or urban entities which have gone beyond the classical physical morphology of the first generation metropolis dominating the twentieth century; beyond the traditional administrative control of local governments; beyond the prevalent sociological reference to inhabitants, with the development of the second and third generation metropolises increasingly dependent on non-resident populations (Martinotti 2004, Mo.Ve 2005).

has resulted in an increase in the times and distances covered by residents to access assets and services and, for more fragile individuals (or lacking mobility capital), to increase the risk of not succeeding in accessing urban opportunities and being excluded from them (Gallez et al. 1997, Grieco et al. 2000, Mignot et al. 2001, Cass et al. 2005).

The results of the most recent research studies carried out on the issue point out that the risk of inaccessibility and social exclusion is particularly high for persons combining lack of local access resources with fragility of individual, family and relational resources (Colleoni 2008). This is the case of low-income families or the elderly and sick, and low-skilled immigrants lacking cultural and relational resources living in peri-urban areas with few services, mainly of the same type, with low-quality facilities and spaces and the absence of alternatives in choosing means of transport. Even if they adopt complex strategies to improve their mobility and accessibility to urban areas, the risk of being excluded from them appears high and likely to grow in contemporary urban societies (Jirón 2007). Although, on average, they are better provided with assets and services than in the past, contemporary urban areas still have many barriers (physical but also spatial-temporal and cultural ones) with the result that they narrow rather than increase the potential for interaction and the space for people's action (Hansen 1959, Dijst 2001).

Survey aims and method

The research belongs to the tradition of studies on the interactions between the spatial and morphological characteristics of cities, the socio-economic structure of the societies living in them and mobility of their populations. Carried out by Mo.Ve in 2005–6 in the metropolitan areas of Barcelona, Bologna, Lyon and Vienna, it aims to analyse the way in which the urban structure of residential areas influences the presence and availability of opportunities and the location of residential areas and opportunities, combined with the different possession of mobility capital by residents, influences styles of mobility and accessibility to urban assets and services (see Figure 7.1).

The data were collected by using two different sources:

- population and housing censuses and local government archives, to find out the presence, numerousness, type and territorial distribution of public assets and services (opportunities);
- a sample survey carried out through time-budget[6] and questionnaire, to find out mobility styles and residents' access.

---

6   The English term 'time-budget' represents the most useful survey method to collect data on the use of time and space of representative samples. For in-depth examination, see Colleoni 2004.

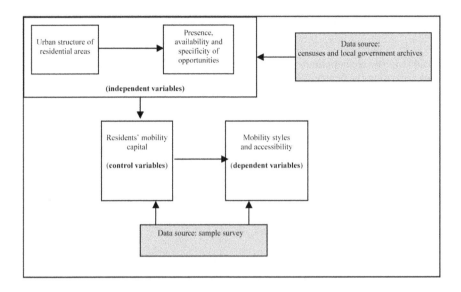

Figure 7.1   Survey hypothesis and data source

The boundaries of the four metropolitan areas were drawn by using the differential density calculation of the resident population[7] and, within them, the municipal areas divided into districts in relation to the spatial distribution of the most significant assets and services and indices of presence of local opportunities. It was thus possible to work out a typology for nine districts according to the position they took on two variables, numerousness and specificity of services present (high, medium and low) and their position regarding the centre of the metropolitan area (central, intermediate and peripheral, see Table 7.1).

The analysis units and survey cases are:

- the territorial district (for a total of 36 districts, nine per metropolitan area);
- the resident nuclear family with children (for a total of 4,000 families, 1,000 per metropolitan area).

---

7   In more detail, the boundaries of the metropolitan areas and, consequently, the local areas surveyed, were identified by starting from the lists of local boroughs included in the metropolitan areas according to the legislation of the four European cities, bringing modifications (inclusion or exclusion of the borough) in cases when demographic density was considerably lower than the average in the area.

Table 7.1   Type of districts

Position in the metropolitan area regarding the centre	Numerousness and specificity of services		
	High	Medium	Low
Central	CH	CM	CL
Intermediate	IH	IM	IL
Peripheral	PH	PM	PL

By independently filling in a questionnaire and a daily time mobility schedule, the adult members of the sample families supplied the necessary data to construct the indices referring to three dimensions:

- quality of individual access resources;
- quality of local access resources;
- mobility styles and access to public assets and services (see Figure 7.2).

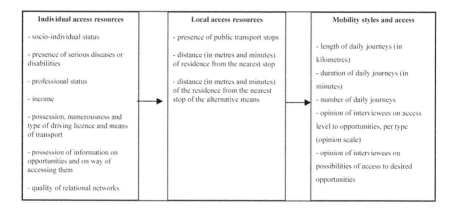

Figure 7.2   Indices for resources and mobility styles and access

Main results

In the four metropolitan areas the transfer of residential, production and business functions from the centre towards the sub and peri-urban areas was followed by the decentralization of many opportunities, formerly accessible only in the central areas. The most frequent peri-urban collocation of urban assets and services and their greater specificity and complementariness had the effect of increasing the average level of inter-dependence between the peripheral areas, increasingly

Table 7.2   Urban concentration index of services per metropolitan area

Metropolitan area	Services			
	Retail Business	Social and Schools	Administration and Health	Leisure
Barcelona	0.39	0.4	0.59	0.7
Bologna	0.32	0.45	0.72	0.85
Lyon	0.27	0.43	0.55	0.7
Vienna	0.44	0.4	0.55	0.49
Total	0.6	0.43	0.61	0.7

interlinked by multi-directional relations compared with the traditional local government units. The effects on the mobility system have been the general increase in mobility behaviours in the sub and peri-urban areas of the city and, above all, between the sub and peri-urban areas in the metropolitan areas.

The research results, however, show that the decentralization of opportunities and the overall increase in mobility have not led to an analogous increase in accessibility; the causes should be sought in the uneven local distribution of services and means to access them in the various metropolitan areas. The greater inter-dependence between peripheral areas has concerned more the places and services of production and, partially, business ones, rather than opportunities in the administrative sector and health and leisure, which continue to find a privileged location in urban centres. The index of urban concentration of services[8] shows the highest average values for leisure services (0.70) and for administration and health (0.61), and the lowest average values for social and school services (0.43) and above all retail business (0.36): see Table 7.2. In other terms, the probability for families living close to opportunities, and therefore to access them over short times and routes, tends to decrease passing from retail services to administrative, health and leisure. Among the four metropolitan areas the best situations are seen in Vienna (with an average concentration index of 0.47) and Lyon (0.49) while the worst are in Barcelona (0.52) and above all Bologna (0.59).

Accessibility to services is also negatively affected by the uneven distribution of means of transport and the location of mobility infrastructures (train/bus/ underground stops, car parks, cycle and pedestrian lanes, etc.). The lack or severe shortage of public transport services brings about the high use of private cars by residents in the outlying districts (59% of cases compared with 45% of residents in semi-central areas and 38% in urban centres, see Table 7.3). The consequences are considerable not only regarding longer travel times but also costs. In the four metropolitan areas the inhabitants in sub-urban districts travel for a longer time

---

8   The index, which measures the level of urban concentration of services, has values between zero (minimal concentration) and one (maximum concentration).

Table 7.3   Indices of mobility per type of urban area

Type of urban area	Average time for journeys (min.)	Average length for journeys (km.)	Use of car (%)	Average number of journeys
Central	36	18	38	2.6
Semi-central	51	22	45	3.5
Sub-urban	72	28	59	4.5
Barcelona	45		28	3.2
Bologna	52		40	3.7
Lyon	59		51	3.7
Vienna	48		36	3.7

and over longer distances, both in reaching work or study places, or for family needs and for leisure activities (respectively 72 minutes and 28 kilometres per day compared with 36 minutes and 18 kilometres for residents in central areas). Moreover, in these districts a greater frequency of journeys is seen, which seems to denote the difficulty in combining several activities in a single journey for those living in more outlying areas (4.5 journeys compared with 2.6 for residents in central areas). For the members of more fragile families, the increased fragmentariness of journeys, which are longer on average, increases the likelihood of having to renounce certain activities if they involve a journey considered too costly, not only in economic terms but also in those of space and time (in particular leisure activities).[9]

Lastly, in some metropolitan areas (above all in Southern Europe), the presence of mobility infrastructures (roads, railway lines, etc.) has had the effect not of decreasing but increasing the isolation of peripheral areas, transforming their spaces from social places to mere transit ones, aimed at facilitating the flow towards the centre rather than in improving accessibility to local services.

Styles of mobility and possibilities for access depend on the location of homes and the local distribution of opportunities combined, however, with the wealth of individual, family and relational resources of the families.

Regarding the issue of individual access resources, the research findings show that the traditional factors of socio-economic differentiation do not act directly on accessibility but through the action brought to bear by the knowledge, ability and practices enacted by the social players in the degree of freedom of choice between possible alternatives.[10] People with more knowledge of urban assets and

---

9   This phenomenon is normally described with the expression *distance decay* (see Colleoni 2008).

10   The Indian economist Amartya Sen effectively synthesizes the set of knowledge, abilities and practices owned by social actors and their degree of freedom of choice between possible alternatives with the respective terms of *functionings* and *capabilities* (see Sen 1993).

Table 7.4   Indices of mobility per family income level

Family income level (net annual in euros)	Income spent on the home and travel (%)	Number of cars per family	Number of cars over eight years old (%)	Average number of means available to reach the city centre
Low (< 11.500)	52%	0.5	34	1.1
Medium (11.500–31.000)	45%	0.6	22	1.5
High (> 31.000)	34%	1.2	11	3.2

services, more skilful in travel practices and with more possibilities for choice between means of transport have greater possibilities for access, according to their socio-economic condition. While diversities in mobility and accessibility between gender, age and income are less marked than in the past, significant differences continue to persist related to the proportion of income devoted to mobility costs, quality of travel and alternatives in choice of means of travel. Dividing the sample into three income level groups (low, medium and high), we may observe that the less well-off families use 52% of their income on the home and travel (compared with 34% of wealthier families), own fewer cars (0.5 per family compared with 1.2) and older cars (34% over eight years old compared with 11% for higher-income families) and, above all, have less freedom of choice between means of transport (1.1 means against 3.2 for high-income families, see Table 7.4). The combination of a fragile socio-economic condition and limited knowledge, ability and freedom of choice forces many families to adopt inadequate styles of mobility to access urban assets and services, particularly if this is added to a disadvantaged local collocation of the home.

The survey findings lastly recall that, apart from being associated with the possibility for choice between various alternatives of means and forms of travel, facility of access also depends on the extension of the space for action, the area providing opportunities which may be reached and used by individuals to carry out their activity (Couclelis 2000, Dijs et al. 2002). The space for action, in particular, shows a greater extension among people with more information, supported by denser relational networks. An effective way to improve the level of accessibility of more fragile persons therefore consists in facilitating the use, rather than ownership, of means of transport (through subsidies for the use of public transport, and above all through the support offered by relationship networks, neighbours and friends) and in increasing their level of knowledge and competence.

Information acts favourably by broadening the perceived space of action, providing people not only with knowledge of the existence and characteristics of the opportunities but also what is necessary for a better learning of the practices to reach places. The possession of knowledge also acts favourably on mobility by reducing the feeling of uncertainty normally associated with unknown places and activities. We must however recall that information and knowledge are only access resources and aspects of mobility capital if, in their turn, they are accessible and heralding capacities and competences for social actors. Otherwise they will inevitably reproduce those very inequalities which, in theory, they should be helping to overcome (Colleoni 2008).

Conclusions

The research has highlighted that, to improve the level of accessibility to urban assets and services and, hence, the level of urban and social integration, it is not sufficient to increase mobility, but it is necessary to improve its quality, by increasing both individual and urban access resources constituting the mobility capital they have. Particular attention must also be addressed to more fragile persons and families, due to their low socio-economic condition and low level of knowledge, ability and freedom of choice and the location of their homes in urban areas which are poor in assets and services. The combination of individual fragility, relational weakness and environmental poverty forces a considerable number of people to adopt inadequate mobility styles to access urban assets and services, exposing them to high risks of urban and social exclusion. The fact that cities today have more assets and services than in the past does not improve the situation, since their distribution in urban areas is still too heterogeneous and their access is made difficult by physical, space-time and cultural barriers which restrict, instead of increasing, people's potential for interaction and space for action.

Where policies exist to support urban accessibility, as in Central and Northern European countries and, regarding the comparison carried out, in the metropolitan areas of Vienna and Lyon, the indices highlight the presence of less fragmentary mobility profiles for the resident populations, less conditioned by the use of a single means of transport (above all private cars). However, the benefits from living in one metropolitan area rather than in another appear less marked than those seen between the different areas inside them, since central areas everywhere best facilitate residents' access to urban assets and facilities.

References

Bagnasco, A. and Le Galès, P. (eds) 2001. *La città nell'Europa contemporanea.* Napoli: Liguori.

Cass, N., Shove, E. and Urry, J. 2005. Social exclusion, mobility and access. *The Sociological Review*, 53, 539–55.

Castells, M. 1996. *The Rise of the Network Society. The Information Age: Economy, Society and Culture*, vol. 1. Oxford: Blackwell Publishing.

Colleoni, M. 2004. *I tempi sociali: teorie e strumenti di analisi.* Roma: Carocci.

Colleoni, M. (ed.) 2008. *La ricerca sulla mobilità urbana: metodo e risultati di indagine.* Milano: Raffaello Cortina.

Couclelis, H. 2000. From sustainable transport to sustainable accessibility: can we avoid a new tragedy of the commons?, in *Information, Place and Cyberspace: Issues in Accessibility*, edited by D.G. Janelle and D.C. Hodge. Berlin: Springer, 341–56.

Dijst, M. 2001. *An Action Space Perspective on the Impact of New Information and Communication Technologies.* Paper to the Sixth Nectar Conference, Helsinki, 16–18 May.

Dijst, M., Schenkel, W. and Thomas, I. (eds) 2002. *Governing Cities on the Move. Functional and Management Perspectives on Transformations of European Urban Infrastructures: Urban and Regional Planning and Development.* Aldershot: Ashgate.

Fouchier, V. 1998. *Urban Density and Mobility in Ile de France Region.* Paper to the Eighth Conference on Urban and Regional Research, Madrid, 8–11 June.

Gallez, C., Orfeuil, J.P. and Polacchini, A. 1997. L'évolution de la mobilité quotidienne: croissance ou réduction des disparités? *Recherche Transport Sécurité*, 56, 27–42.

Grieco, M., Turner, J. and Hine, J. 2000. *Transport, Employment and Social Exclusion: Changing the Contours through Information Technology.* [online] Available at: http://www.geocities.com/transport_and_society/newvision.html [accessed: 26 February 2009].

Hansen, W.G. 1959. How accessibility shapes land-use. *Journal of the American Planning Institute*, 25, 73–6.

Jirón, P. 2007. *Mobile Place-making in Santiago de Chile. The Experience of Place Confinement and Place Autonomy.* Paper to the 38th World Congress of the International Institute of Sociology, Budapest, 26–30 June 2008.

Kaufmann, V., Bergman, M.M. and Joye, D. 2004. Mobility as capital. *International Journal of Urban and Regional Research*, 28 (4), 745–56.

Martinotti, G. 1999. Introduzione, in *La dimensione metropolitana: Sviluppo e governo della nuova città*, edited by G. Martinotti. Bologna: il Mulino.

Martinotti, G. 2004. *The Rise of Meta-Cities: Mobility and the New Metropolitan Europe.* Paper to the 3th Forum of Mo.Ve (International, non Governmental, Permanent, Observatory on Sustainable Mobility in Metropolitan Areas), Venezia 21–2 October 2004.

Mignot, D. et al. 2001. *Mobilitè et grande pauvreté*. Final Report, Agence d'urbanisme pour le développement de l'agglomération lyonnaise – Observatoire Social de Lyon.

Mogridge, M.H.J. 1985. Transport, land use of energy interaction. *Urban Studies*, 22, 481–92.

Mo.Ve (International, non Governmental, Permanent, Observatory on Sustainable Mobility in Metropolitan Areas). 2005. *Final Report*, 4th Forum, Venezia, 29–30 September 2005.

Naess, P. 2006. Accessibility, activity participation and location of activities: exploring the links between residential location and travel behaviour. *Urban Studies*, 43 (3), 627–52.

Naess, P., Gunnarroe, P. and Larsen, S. 1995. Travelling distances, modal split and transportation energy in thirty residential areas in Oslo. *Journal of Environmental Planning and Management*, 38, 349–70.

Naess, P. and Jensen, O.O. 2004. Urban structure matters, even in a small town. *Journal of Environmental Planning and Management*, 47, 35–57.

Newman, P.W.G. and Kenworthy, J.R. 1989. *Cities and Automobile Dependence*. Aldershot: Gower Publications.

Sassen, S. 2001. *Global Cities*. Princeton: Princeton University Press.

Schwanen, T., Dieleman, F.M. and Dijst, M. 2001. Travel behaviour in Dutch monocentric and polycentric urban system. *Journal of Transport Geography*, 9, 173–86.

Sen, A. 1993. Capabilities and well-being, in *The Quality of Life*, edited by M. Nussbaum and A. Sen. Oxford: Clarendon Press, 30–53.

Stead, D. and Marshall, S. 2001. The relationships between urban form and travel patterns: an international review and evaluation. *European Journal of Transport and Infrastructure Research*, 1, 113–41.

Urry, J. 2002. Mobility and proximity. *Sociology*, 36 (2), 255–74.

Chapter 8

Mobility Practices in Santiago de Chile: The Consequences of Restricted Urban Accessibility

Paola Jirón

Differentiated mobility refers to the diverse ways people experience urban daily mobility according to gender, life cycle, religion, income, age, ethnicity, or ability, amongst others. Moreover, because social practices remain based on uneven power relations, social differences may exacerbate them and impact the possibilities of accessibility to people, activities and places. This means that for some people, their social characteristics provide them with open passports to access all sorts of domains in urban areas, a form of 'laisser passer' through the city. For others, their social conditions limit their connection, flow and accessibility, leaving them in longer queues, with restricted access and limited possibilities. Under a mobility lens, urban inequality refers to uneven access to practices, relations and places leading to temporary or permanent connections or disconnections in timespace. Therefore, different social conditions expressed in daily mobility practices combined with daily mobility barriers generate a complex web of relations taking place in cities today. However, little is known about the way these relations take place, and this chapter provides further detail in the case of specific residents in Santiago de Chile.[1]

Access has been recognized as a major aspect of social exclusion (Cass et al. 2005, Kenyon et al. 2003), and increase in transport is often seen as the main possible solution. A closer look at the practices of daily mobility and the way inequality is experienced reveals that transport can be a major barrier, but that there are also other barriers that need to be overcome. This involves seeing social exclusion as a process as opposed to outcomes. In closer detail and from an urban daily mobility approach, transport systems can sometimes be distinguished as deficient but not necessarily the main barrier of exclusion. This is because mobility entails more than travelling from point A to point B; it involves understanding what occurs in mobility practices, how they occur and what happens prior to and following the practice. Mobility can sometimes be the cause and other times a consequence of uneven social relations, or the manifestation of more profound

---

1   Parts of this chapter are based on FONDECYT financed research N° 1090198.

inequalities in urban living. Consequently, within everyday practices of daily mobility, social exclusion can be analysed through the concept of accessibility.

When understood as level of connectivity, Church et al. (2000) point out that accessibility is only one dimension of social exclusion,[2] and high accessibility does not imply people are able to benefit from it (Church et al. 2000). Thus, to understand how mobility affects social exclusion, adapting Cass et al.'s work (2005), accessibility here is understood as the ability to negotiate space and time to accomplish daily practices, maintain relations and generate the places that people require for social participation. Although it does not capture all the dimensions of social exclusion, this definition provides a deeper comprehension of the implications of being connected or disconnected, of the capacities people have to enter or exit, the consequences of being left out or choosing to stay out or in, thus looking at the types of connections, the times, places and relations.

This chapter argues that urban daily mobility practices are differentiated according to social conditions of gender, income, age, stage in life cycle, amongst others, and this differentiation affects people's accessibility to various aspects of daily living. These differences are enhanced when physical, financial, organizational, temporal, technological and skills-related dimensions of mobility restrict access to practices, relations and places, becoming mobility barriers and generating experiences of inequality. This chapter briefly introduces the discussion on urban inequality from a mobility point of view. It then provides a detailed description of individual daily trajectories in Santiago in terms of access to the specific practices of going to work, analysing how the specific mobility barriers unveil inequality issues related to gender, household responsibilities, income, technology, time and flexibility.

Access to practices, relations and places

The literature on mobility, mainly from transport studies in Europe and the USA provides various measures for accessibility (Miller 1999, Baradaran and Ramjerdi 2001, Hine and Mitchell 2001, Kenyon et al. 2002, Hine and Grieco 2003, Kenyon et al. 2003, Miller 2005a, 2005b, Kenyon 2006, Miller 2006). As such, it is seen as the most 'prevailing measure used by planners and politicians to bolster their everyday propositions' (Baradaran and Ramjerdi 2001: 32). Although most authors agree that there is no universally acknowledged measure of accessibility, and it is often understood in a straightforward way as connectivity either from the supply side or the demand side. Miller (1999) classifies three types of approaches:

---

2   Kenyon et al. (2002, 2003) suggest nine dimensions of social exclusion that influence lack of mobility: economic, living space, mobility, organized political, personal, personal political, social networks, societal and temporal.

(i) constraints-oriented approach,[3] (ii) attraction accessibility measures[4] and (iii) transport benefits, usually seen as utility maximization[5] measures.[6] Most transport models use the latter, calculated in terms of cost benefit, using data intensive models, and measuring it in monetary units of cost. However, according to Axhausen et al. (2002), recent research into the processes by which travellers allocate their time has clearly revealed that this most widely used paradigm, utility maximization, is incomplete in its lack of understanding of the rhythms, routines and habits that make up daily life.

Within the area of transport, travel behaviour research has greatly advanced since Hansons' work (Hanson and Hanson 1980, 1981) which clearly linked travel behaviour with daily travel activity and conceptualized travel as more complex than simple moving from A to B. As suggested by Law (1999), a considerable amount of work has been carried out to explain the relation between travel and gender inequality[7] as well as disability and transport. Today, the analysis of accessibility is developed mainly through quantitative data, modelling daily life and mobility (Axhausen et al. 2002, Kwan et al. 2003, Kwan and Lee 2003, Miller 2005a, 2005b, 2006, Ohnmacht 2006) using mostly the gravity and opportunities approach based on spatial opportunities available to travellers. Although these analyses are necessary and useful to understand aggregate patterns of travel behaviour, they still lack the understanding of the specificity of the experience of travelling for many groups of people, how it impacts their access to urban benefits and how this practice relates to other aspects of urban living. Therefore, travel behaviour studies would be significantly enhanced if complemented with more detailed research that explored the experiences of daily mobility.

Within the broader social policies literature, accessibility is becoming a key issue in the discussion of inequality and exclusion (Church et al. 2000, Baradaran and Ramjerdi 2001, Hine and Grieco 2003, Kenyon et al. 2003, Lyons 2003, Schönfelder and Axhausen 2003, Cass et al. 2005, Miller 2006). Specifically in the UK, the Social Exclusion Unit (SEU 2003) has defined it as the way people access key services at reasonable cost, in reasonable time and with reasonable ease. Thus accessibility analysis involves not just approaching transport but also the location and delivery of key activities. In this type of analysis, adequate access

---

3   Based on Hägerstrand (1970) time and space constraints.

4   Based on spatial opportunities available to travellers (Baradaran and Ramjerdi 2001) also known as the gravity and opportunity approach.

5   Based on travel demand modelling, it depends on the groups of alternatives being evaluated and the individual (Baradaran and Ramjerdi 2001) and is measured in monetary units.

6   Baradaran and Ramjerdi (2001) complement these with travel cost (ease with which any land-use activity can be reached from a location using a particular transport system) and composite approaches (which Miller (1999) presents as a composition of constraints and utility based).

7   For a detailed evolution of the study of transport and gender in Western societies see Law (1999).

would involve knowledge of transport as well as the experience of it, trust in its reliability, and having physical and financial access to it.[8]

This way of looking at accessibility has become extremely useful to acknowledge mobility implications of social exclusion, yet it presents some problems, as it is presented as a somewhat top down approach with little consideration of actual practices individuals carry out daily or the way people use mobility for purposes other than transport. Furthermore, it says little about the way people access the network of relations they have, as suggested by Cass et al. (2005). For this, Hine and Grieco (2003) suggest distinguishing between direct and indirect accessibility, where the first refers to 'ability of individuals to plan and undertake journeys by public or private modes subject to time budget and cost' (Hine and Grieco 2003: 300), whereas the second refers to 'the extent to which individuals or groups can rely on neighbours or other support networks to access goods and facilities on their behalf subject to time and financial budgets' (Hine and Grieco 2003: 300). This implies that, a transport-only approach limits the comprehension of inequality in mobility experiences.[9]

Consequently, accessibility analysis here is based on Cass et al.'s approach to access as 'the ability to negotiate space and time to accomplish practices and maintain relations that people take to be necessary for normal social participation' (Cass et al. 2005: 543). The relevant aspect of this definition is negotiation, as it is in the daily intersections in timespace that connection and disconnection are more likely to take place, become a problem or an opportunity and manifest existing inequalities. For this research, however, this definition is expanded to include the way in which individuals and groups negotiate access to practices, relations *and* places. Place has been added given the importance of mobile place making (Jirón 2010a). Thus access to mobile places becomes an important aspect in urban inequality under the mobility lens, and this, along with access to relations.

In the process of negotiating access, the various ways in which it may be restricted, – for instance by social, physical, economic, or even environmental limitations – indicate that inaccessibility may be related to uneven location of infrastructure or inadequate public transport systems.[10] However, uneven access

---

8   SEU report Transport and Social Exclusion (February 2003) led to Accessibility Planning becoming embedded in the work of Local Authorities. Accessibility Planning seeks to ensure that there is a clear and more systematic process for identifying and tackling barriers that people face in accessing jobs and key services such as education and healthcare (SETF 2007).

9   Kenyon et al.'s (2003) work is relevant in highlighting the need to look at mobility from a broader point of view to include virtual mobility and not just physical one, i.e. transport.

10   Existing infrastructure affecting frequency, quality and availability of public transport could be seen as a physical boundary inhibiting mobility. During this research, some of the problems detected in Santiago's public transport system included overcrowded, unsafe, insecure and unreliable buses, inconvenient routes, unsafe bus stops, rude drivers,

may also stem from factors such as existing uneven gender relations within the household or society, or cultural barriers that prevent different groups from mixing or encountering each other, amongst others. This implies that social differences such as gender, age, income, ability, religion or ethnicity, may generate differentiated experiences of mobility, which could lead to restricted accessibility. Thus, improving accessibility involves thinking about factors beyond the elimination of physical barriers or the creation of infrastructure, services or housing, as perhaps more transport may have perverse consequences for social exclusion (Shove 2002). It also refers to the need to observe the capacity and possibility of making use of such opportunities in terms of motility.[11] Therefore, in this research, both aspects of mobility barriers are relevant: the persons' motility and the existing structures that constrain or enable mobility.

Time geography pioneered in the 1970s and introduced the indissoluble link between time and space. In his elaboration of time space mapping, Hägerstrand (1970) developed the two major constraints to accessibility: time and space, through three types of time space constraints: coupling, capability and authority constraints.[12] However, time and space do not capture fully the complexity of the barriers present in mobility. Church et al. (2000) identified seven dimensions that act as barriers to accessibility: physical, geographical, activities and facilities, economic, time, fear and space. In turn, Cass et al. (2005) have synthesized these into four key dimensions of access: financial, physical, organizational and temporal. Law (1999) also presents skills and technology as mobility barriers.

For this research, accessibility to practices, relations and places is observed according to financial, physical, organizational, temporal, skills and technological barriers. Financial barriers affecting mobility may involve, for instance, the cost of using different modes of transport; physical dimensions may relate to the distance travelled but also the physical aspect or condition of spaces encountered, including roads, sidewalks, bus stops, buses, metros, platforms, bike paths, or

---

and cost of public transport. Many of these issues were part of the diagnosis used to implement a major transformation of Santiago's transport system: Transantiago. The new transport system was in its pilot phase during the time the research was conducted and was implemented fully in February 2007. Although public transport in Santiago has changed since fieldwork was undertaken, many of these mobility issues persist and, in many cases, have been exacerbated.

11   Motility refers to the process in which 'an individual or group takes possession of the realm of possibilities for mobility and build on it to develop personal projects not necessarily transforming it to travel' (Flamm et al. 2006: 168).

12   Coupling constraints define where, when, and for how long individuals can join other individuals, tools and materials in order to produce, consume and transact; capability constraints are those which limit the activities of individuals due to physical (distance) or biological (sleeping, eating) factors; authority constraints refer to a domain or control area where things and events are under the control of certain individuals or groups that set limits on access (requiring payment, invitation, ceremony, fight). These three aggregations of constraints interact (Hägerstrand 1970).

parks, amongst others. Organizational restrictions deal with the multiple activities people carry out on a regular basis in order to coordinate daily living, including shopping, attending health facilities, paying bills, accessing work, etc. Temporal dimensions involve looking at the way day, night, seasons, opening hours, and duration of trips affect mobility decisions. Skills refer to the capacities people have to be able to move in specific ways, for instance knowing how to drive or having a drivers licence, knowing how to change tyres, knowing how to cycle or being able to fix a bicycle, being able and feeling comfortable about riding a motorbike. Technological barriers involve the possibility, capability to use and availability of technology to enhance or facilitate journeys or substitute the need for physical travel, including the Internet or mobile phones. Each of these barriers is altered when looked through individual and group socio cultural conditions, including gender, age, stage of lifecycle, ability or ethnicity.

Sociocultural characteristics are seen as an additional barrier that influence all the above-mentioned mobility barriers and greatly impact the differentiated way mobility is experienced. This set of barriers is presented in Figure 8.1. Inequality in Mobility Framework, to analyse mobility practices, where, for instance, access to travel might be physically impaired if bus stops are not located in convenient sites, but this complication is enhanced for the elderly who have extra difficulty walking, causing them to minimize their use or stop using the specific mode of transport. Similarly, transport availability at specific times may be an issue for most travellers, but as a barrier it becomes particularly difficult when women fear for their own security when travelling alone at night.

Accessibility in itself is neither good nor bad; its evaluation depends on the implications it has on everyday life. Unevenness in accessibility becomes a problem when it is not voluntary, when people lack alternatives, when it is mandatory, when the only options are to remain disconnected or perpetuate unwanted connections. It becomes a problem when greater possession of capitals provides greater disproportionate access, or when structures in society enhance differences, leaving people outside of the benefits of society. Analysing accessibility and how it may lead to involuntary connection or disconnection and uneven social relations within the practice of mobility involves looking at the strategies people use to access and overcome existing barriers. In the strategies for coping, challenging,

Mobility Barriers/ Everyday Accessibility Dimensions	Socio-cultural Conditions					
	Physical	Financial	Organisational	Temporal	Skills	Technology
Practices						
Relations						
Places						

Figure 8.1   Inequality in mobility framework

defying or transgressing existing accessibility barriers, the difficulties households and individuals face are envisaged, but also in these strategies the possibility to negotiate, encounter, exchange or create something completely new can be found. Although many strategies are individual, more often than not, they have household implications, particularly in terms of routine organization or use of time, and they often involve third parties like friends or family. Strategies here will be understood as the set of practices that are implemented to improve or maintain access to timespace, while keeping or improving the volume of capital.

These practices have specific implications in terms of uneven experiences of mobility, where at times socio-cultural conditions are in themselves the major barrier generating such unevenness, while other times uneven access can be attributed to physical, financial, temporal, organizational, technical or skill boundaries that impact different groups in uneven manners. People devise multiple strategies to enable mobility, and perhaps immobility is not the main problem for many, but rather the hardships experienced during mobility practices and the meaning given to these experiences.

The following sections present the results from an ethnographic study carried out in Santiago de Chile, where participants were shadowed during their daily mobility routines.[13] The first section compares the experiences of two men with different mobility issues which impact on their sense being confined. The next section presents a different type of experience, by young people on the move. Both cases provide insights on the way mobile place confinement and enlargement occur in cities like Santiago today and the possibilities of shifting them.

Roberto and Francisco

Education is a barrier for many urban workers, for some it implies making important sacrifices in terms of the distances travelled and the time spent at work. For other, their limited access to education and home responsibilities only allow them to access certain types of jobs with certain characteristics as will be seen through the cases of Roberto and Francisco.

Roberto

Roberto is 42 years old, married to Cecilia and together they live in an upper-middle-class neighbourhood. At the age of 25 he married and soon after separated. Unlike his wife Cecilia, Roberto never went to University, he is an accountant by trade and his limited access to education greatly hinders his chances of more stable job and better salary. He makes half as much as his wife for child maintenance,

---

13   The research compared mobility practices of residents living in three different income neighbourhoods located relatively close to one another. For a further detail on the methodology used see Jirón 2007a, 2007b, 2008 and 2010a, 2010b.

Figure 8.2   Roberto walks to *colectivo* stop

out of which he gives about a third to his ex-wife. In his current job, he acts as the accountant and hopes that after five years there he will be able to get a better job as a proper accountant. He is not overly excited about his job and he works very long hours, including Saturdays. Cecilia demands more time and motivates him to study, but his daily routine does not really allow it. He works in the other side of the city, on the West, close to the airport. The couple only have one car which she uses to get to work, they cannot afford a second one for now. With the newly built highway, this trip would take him about 25 minutes with normal traffic but with the *colectivo*[14]/metro/bus ride, it takes him two hours.

His journey starts at 6.40a.m. when he leaves his house walking towards the *colectivo* stand (see Figure 8.2). At this time the queue is short, so he is soon on his way. The drive to Mirador Station is quick and by 7.00a.m. he's already on the metro. By 7.20a.m., after 10 stops he changes line on Baquedano Station, towards the west for 13 stops, rush hour is just starting, so the platforms and wagons are fuller, particularly with people going downtown (see Figure 8.3, Roberto inside the Metro). By 8.00a.m. he's at his final Metro Stop in Pajaritos. While he waits for the bus that will take him to his job, he buys breakfast. At 8.10a.m., the bus which takes him to ENEA, an industrial centre next to the airport, where mainly

---

14   A fixed rate shared taxi.

Figure 8.3   Roberto inside the Metro

offices and warehouses are located, and by 8.40a.m. he is in his office, and can have breakfast before everyone else arrives. After 9.00a.m. the buses towards and from ENEA reduce their frequency, so there is limited chances of leaving the area unless by car, and the surrounding area is bare and under construction (Figure 8.4, Roberto waits at the bus stop). At 7.30p.m. he is ready to leave his work and walks to the bus stop where other employees from various companies are already waiting. He repeats the same journey back, gets back to the Metro which is still very full, changes at Baquedano, and takes the metro until Mirador Station where he queues for his *colectivo* and gets off a few blocks away from home. By the time he gets home at 9.30p.m., his wife is already there, waiting and ready to warm up some food for him. He gets ready for the next day.

The trips Roberto makes during the day make it difficult for him to actually encounter other people, although the distance travelled is far and the *colectivos* and metro is full, his mind drifts away and is immersed in self reflection, he constantly looks outside, at times enjoying the ride but mostly wishing to be elsewhere. His strategy is along the lines of putting up with the job he has and the travelling situation for the next five years, he is therefore immobile in his mobility.

Figure 8.4   Roberto waits at the bus stop

Francisco

Francisco is 42, married to Alejandra and lives in Santa Teresa, a low-income neighbourhood. They have two children, Sandra and Sergio who are 18 and six respectively. They both come from the Concepcion area in the South of Chile, and have been in Santiago for more than 20 years each. He did not finish high school and does not have any technical training. Upon arriving to Santiago he moved around in various jobs, including butcher shops, supermarkets and construction sites and has now been working as a security guard at a very high income gated community close by. He chose this job because his son Sergio has a learning disability, therefore he needs to go to the doctor often, although he goes to school for a few hours a day, he needs constant attention. Francisco works on the night shift and takes care of his son during the day. He says he makes less money than in other jobs he could have, but then he would have to pay for someone to take care of Sergio and the boy would not receive his personalized attention.

He works five nights from 10.00p.m. until 8.00a.m. for 5 days then he gets two days off. Prior to going to work he prepares his lunch box, which usually involves left over dinner, a few sandwiches and a thermos of tea. His route is short, takes him about 20 minutes through the park, a football field, some empty lots, a shanty town and then formal housing (see Figure 8.5) The gated community he works at

Figure 8.5   Francisco rides bicycle to work

covers a large area; it can be entered from one street and exited on the next. When he comes back in the morning he sleeps until around 11.00a.m., prepares Sergio, feeds him, makes him do his homework and takes him to school at 2.00p.m. on his bicycle to a school relatively close by, about a 20 minutes ride (Figure 8.6). When he comes back he tries to sleep again from 3.00p.m. to 5.00p.m. when he has to go on the bicycle to pick up Sergio who gets out at 5.30p.m. He then prepares his tea time, waters the garden, cleans the house (Figure 8.7) until his wife Alejandra gets home around 7.30p.m., and he goes back to sleep until about 9.30p.m. when he gets ready for work again.

He says he likes riding the bicycle, but it tires him out too taking him there and back, it's two trips plus his weight, 'sometimes I go to the 14th, or the street market on the bicycle, it's tiring … But I don't take buses or *colectivos*, unless it's necessary … It's cold at times riding, but I don't really have much choice because I don't have buses going there, I would have to take a *colectivo*, but they are not available at night. So I would have to walk and it could take me over an hour' (Francisco).

His routine is tiring he says, 'it's like I do two shifts, because in the morning I have to watch Sergio, take him to school' (Francisco). This pace is exhausting, and the lack of sleep and physical hardship raises questions about the sustainability of this practice, as Sergio gets bigger and Francisco older.

Figure 8.6   Francisco takes Sergio to school

The strategy Francisco's family uses for coping with their child who needs a lot of attention is having one of the parents constantly with him. Alejandra also works cleaning a house close to where he lives, but walking is too far and dangerous for her, so she takes a *colectivo*. She chose to work here as a strategy to be close to home in case of an emergency. The area where she works is very dangerous at night, so she tries coming home early, but it is not always easy. The rigidity of Francisco's job (fixed hours of required presence at night) and Alejandra's daily job, allows this household to be able to take turns taking care of their child. Any other situation would require them to pay for a service that they cannot easily afford. This also allows Francisco to be able to drop him off at school everyday, talk to the teachers, and take him to the doctors as often as is required.

In Roberto's case, he does not have the education level to get a better job, or to the salary to buy himself a car which would make his journey easier, therefore he has to work for long hours in a very distant job, which he does not really enjoy and spends approximately four hours of his day travelling. His possibilities for improving his situation, getting better skills, could be through studying, but given the way night courses work, the times and distances he is required to travel would not allow him to do it and it would be practically impossible given his Saturday shifts, This would also mean even later arrival at home. His strategy is sticking it out for five more years. Roberto is completely disconnected from any possibility

Figure 8.7   Francisco at home

available in the city, since they moved to the new house, he has not been able to enjoy it very much, they do not go out, he barely has time for shopping, or going to *la feria*[15] as he used to, or improving his abilities.

Both men are confined in their mobility experiences. Although one crosses the city daily he is confined on the *colectivo*/metro/bus ride, he looks out the window, but real possibilities of accessing new places are limited. Francisco has to limit the distance travelled in order to make it with his bicycle and be able to carry his son around; he also forgoes a possible better job, in order to be accessible to his child. He is confined within a limited area in the city.

Catalina and Rodrigo

Wanting to do better in life is a shared aspiration amongst those interviewed with children. When asked about what they aspire to in the future, most parents mention their children's success: 'that they become more than me' (Ana). Although the road to a better future is difficult for all, regardless of income, the difficulty of

---

15   Street market.

accessing opportunities vary among various income groups, particularly for the young but also in terms of gender.

Catalina

Catalina is 19; she managed to get a bank loan to go to university to study design, with her father as a bank guarantor. To get the loan she had to push her parents to formalize their separation as her mother had high levels of debt that didn't allow

Figure 8.8   Catalina waits at the bus stop

Figure 8.9   Catalina inside the bus

her to become subject of credit. She knows she is having the opportunity of her lifetime, and she's making the best of it. Every evening, before going to bed, she prepares her bags for the next day, which starts around 8.00a.m. She walks for about 20 minutes to catch the bus (Figure 8.8), this is the fastest and most direct route she has found as she noticed that 'the earlier I get up, the later I arrive, because of traffic and the queues' She also saves money, as she only pays $240[16] daily on the bus. She struggles on the bus, as many go by before she can actually catch one. Bus drivers don't like students; they pay for half price and take up more room with their big bags. Hers is no exception, and she feels uncomfortable with it, but has to take her materials to school. The ride to school lasts about half an hour, all of it standing and balancing along the way, getting pushed, shoved, squished, fondled, stepped on, stared and yelled at (Figure 8.9). By the time she gets off, she's exhausted. But going there is not half as bad as coming back, the same pushing and shoving, but the experience worsens as she tries to get off, the bus driver does not stop for her, and she has to get off two stops later, after getting through the mass of people, mostly men that take the opportunity to touch her as she squeezes through. She feels embarrassed, scared and frustrated. She then walks home, the walk is up hill, but she takes the same amount of time because she walks fast, it is dark already and the streets are scary for her, she is terrified of stray dogs, and they are all over. She gets home around 7.00p.m. She carries out this routine three times a week, the rest of the time she stays home studying.

She will have to develop strategies to deal with this discomfort, either travelling at different times, with friends or using different modes, but not all these options are suitable for her schedule or budget. She then would either have to limit her places of circulation or become more aggressive in her travel behaviour.

Rodrigo

Rodrigo is also 19, and wants to be a builder to have better opportunities in life. His father Bernardo works in a printing shop and his mother stays at home, occasionally sewing clothes for others, so they cannot help him out with money, but he works in a construction site as a plot outliner and parallel to this he studies at night in an institute downtown. His days start at 6.45a.m. and he walks to work, for about half an hour. He works there until 6.00p.m. but has to make it to the Institute by 7.20p.m. He quickly showers and sets off. He takes a *colectivo* to the Metro that takes him downtown. All in all, it takes him over an hour to get there. The ride is difficult as at that time, traffic going to the Metro is heavy and the Metro itself is full. The hardest moments are when he has to change lines. He arrives a few minutes late, but others do as well.

Classes finish at 10.30p.m., and he does not always make it to the Metro before it shuts down, so he has to take a bus back. The bus does not come often, but at least it is only one bus. By the time he catches it, it is after 11.00p.m., and by the time he gets home, it is after midnight. Daily he spends about $1,500.[17] He is exhausted and his mother leaves food on the table for him. He eats and goes to sleep, to get up at 6.45a.m. again, and repeat the same routine five days a week.

---

16   Approximately 24p.
17   Approximately £1.50.

Both mobility experiences are difficult, not only because the public transport system is not adequate to the needs of students or less to those working and studying but also due to gender differences. Catalina finds it very difficult travelling on the bus, because of the treatment from the driver and other passengers but also because she does not feel safe travelling on her own. Unlike Rodrigo, she would never travel at night, scared of being mugged or harassed. She is making it work for her though, and she understands that soon she will have to become more aggressive in her behaviour on the bus. Rodrigo feels no threat and travelling at midnight is not a problem for him, whereas Catalina feels unsafe at night and in general, prefers not going out at night. Rodrigo does not see a way out soon, it is only two years he says; his only concern is whether his girlfriend will understand that he does not have time for her. His mobility possibilities are expanding, both physically and educationally.

Conclusions

Issues of access to the benefits being produced by societies such as the Chilean one, require bearing in mind how people use the city instead on imposing ways of using it. The details of the experiences of differentiated mobility in the city are not well known or researched yet, and even less incorporated into urban, transport or housing policy, yet these have daily consequences on people's lives. The differentiated experience of mobility can shed light, first of all, on the need for a better transport system in terms of being affordable, available, accessible, acceptable, comfortable, and safe. It can also lead to better connected infrastructure, improved housing interventions, but also the need to improve access to better working conditions, educational and health services, cultural activities, use of leisure time, and the recognition of the informal economy in operation.

The issues presented in this chapter need to be dealt with within broader social policies than transport ones, and in this context, transport policy must be integrated to social policy. Although public transport accessibility will undoubtedly improve mobility inequalities experienced by some groups and individuals, these inequalities need to be looked at more explicitly. This is because although they are transport-related they are also linked to other areas of daily living including employment, recreation, childcare, household relations, amongst others.

References

Axhausen, K.W., Zimmermann, A., Schönfelder, S., Rindfüser, G. and Haupt, T. 2002. Observing the rhythms of daily life: a six-week travel diary. *Transportation*, 29(2), 95–124.

Baradaran, S. and Ramjerdi, F. 2001. Performance of accessibility measures in Europe. *Journal of Transportation and Statistics*, September/December, 31–48.

Cass, N., Shove, E. and Urry, J. 2005. Social exclusion, mobility and access. *The Sociological Review*, 53(3), 539–55.

Church, A., Frost, M. and Sullivan, K. 2000. Transport and social exclusion in London. *Transport Policy*, 7(3), 195–205.

Flamm, M. and Kaufmann, V. 2006. Operationalising the concept of motility: a qualitative study. *Mobilities*, 1(2), 167–89.

Hägerstrand, T. 1970. What about people in regional science? *Papers in Regional Science*, 24(1), 6–21.

Hanson, S. and Hanson, P. 1980. Gender and urban activity patterns in Uppsala, Sweden. *Geographical Review*, 70(3), 291–9.

Hanson, S. and Hanson, P. 1981. The travel-activity patterns of urban residents: dimension and relationships to sociodemographic characteristics. *Economic Geography*, 57(4), 332–47.

Hine, J. and Grieco, M. 2003. Scatters and clusters in time and space: implications for delivering integrated and inclusive transport. *Transport Policy*, 10, 299–306.

Hine, J. and Mitchell, F. 2001. Better for everyone? Travel experiences and transport exclusion. *Urban Studies*, 38(2), 319–32.

Jirón, P. 2007a. Unravelling invisible inequalities in the city through urban daily mobility. The case of Santiago de Chile. *Swiss Journal of Sociology*, 33(1) Special Issue on Space, Mobility and Inequality, 45–68.

Jirón, P. 2007b. Place making in the context of urban daily mobility practices: actualising time space mapping as a useful methodological tool, in *Sensi/able Spaces – Space, Art and the Environment*, edited by E. Huijbens. Cambridge: Cambridge Scholars Press.

Jirón, P. 2008. Mobility on the move: examining urban daily mobility practices in Santiago de Chile. London: London School of Economics and Political Science, Department of Geography and Environment. PhD in Urban and Regional Planning, 385.

Jirón, P. 2010a. Repetition and difference: rhythms and mobile place-making in Santiago de Chile, in *Geographies of Rhythm, Nature, Place, Mobilities and Bodies*, edited by T. Edensor. London: Ashgate, 129–43.

Jirón, P. 2010b. On becoming la sombra/the shadow, in *Mobile Methods*, edited by M. Büscher, J. Urry and K. Witchger. London: Routledge, 67–99.

Kenyon, S. 2006a. The 'accessibility diary': discussing a new methodological approach to understand the impact of Internet use upon personal travel and activity participation. *Journal of Transport Geography*, 14, 123–34.

Kenyon, S. 2006b. Reshaping patterns of mobility and exclusion? The impact of virtual mobility upon accessibility, mobility and social exclusion, in *Mobile Technologies of the City*, edited by M. Sheller and J. Urry. Abingdon: Routledge, 102–20.

Kenyon, S., Lyons, G. and Rafferty, J. 2002. Transport and social exclusion: investigating the possibility of promoting inclusion through virtual mobility. *Journal of Transport Geography*, 10(3), 207–19.

Kenyon, S., Lyons, G. and Rafferty, J. 2003. Social exclusion and transport in the UK: a role for virtual accessibility in the alleviation of mobility-related social exclusion. *Journal of Social Policy*, 32(3), 317–38.

Kwan, M.P., Janelle, D.G. and Goodchild, M.F. 2003. Accessibility in space and time: a theme in spatiality integrated social science. *Journal of Geographical Systems*, 5(1), 1–4.

Kwan, M.P. and Lee, J. 2003. Geovisualization of human activity patterns using 3D GIS: a time-geographic approach, in *Spatially Integrated Social Science: Examples in Best Practice*, edited by M.F. Goodchild and D.G. Janelle. New York: Oxford University Press.

Law, R. 1999. Beyond 'women and transport': towards new geographies of gender and daily mobility. *Progress in Human Geography*, 23(4), 567–88.

Lyons, G. 2003. The introduction of social exclusion into the field of travel behaviour. *Transport Policy*, 10(4), 339–42.

Miller, H.J. 1999. Measuring space-time accessibility benefits within transportation networks: basic theory and computational procedures. *Geographical Analysis*, 31, 187–212.

Miller, H.J. 2005a. A measurement theory for time geography. *Geographical Analysis*, 37(1), 17–45.

Miller, H.J. 2005b. Place-based versus people-based accessibility, in *Access to Destinations*, edited by D. Levinson and K.J. Krizek. Oxford: Elsevier, 63–89.

Miller, H.J. 2006. Social exclusion in space and time, in *Moving through Nets: The Physical and Social Dimensions of Travel. Selected Papers from the 10th International Conference of Travel Behaviour Research*, edited by K.W. Axhausen. London: Elsevier, 353–80.

Ohnmacht, T. 2006. Mapping Social Networks in Time and Space. *Arbeutsberichte Verkehr und Raumplanung*. Zurich: IVT, ETH Zurich, 33.

Schönfelder, S. and Axhausen, K.W. 2003. Activity spaces: measures of social exclusion? *Transport Policy*, 10(4), 273–86.

SETF. 2007. Context for social exclusion work. [online] Available at: http://www.cabinetoffice.gov.uk/social_exclusion_task_force/context/ [accessed: 7 August 2008].

Shove, E. 2002. Rushing around: coordination, mobility and inequality. *ESRC Mobile Network Meeting*, London, October.

Index

Milton Keynes UK
Ingram Content Group UK Ltd.
UKHW031133141024
449569UK00006B/225